Quantifying and Modeling Soil Structure Dynamics

Advances in Agricultural Systems Modeling
Transdisciplinary Research, Synthesis, and Applications
Laj Ahuja, Series Editor

Volume 3

Quantifying and Modeling Soil Structure Dynamics

Sally Logsdon, Markus Berli, and Rainer Horn, Editors

Soil Science Society of America, Inc.
5585 Guilford Road, Madison, WI 53711-5801 USA

soils.org
dl.sciencesocieties.org
SocietyStore.org

ISBN: 978-0-89118-956-5 (print)
ISBN: 978-0-89118-957-2 (electronic)
doi:10.2134/advagricsystmodel3
Library of Congress Control Number: 2013933711

Advances in Agricultural Systems Modeling
ISSN: 2163-2790 (online)
ISSN: 2163-2774 (print)

Cover design: Patricia Scullion

Printed in the United States of America.

Contents

Foreword

Soil structure is often considered as a static property in descriptions and interpretations of soils in the field. However, this is not the case. Soil structure is dynamic. It changes due to a multitude of factors and time scales. These changes directly affect soil properties such as solute transport, root growth and bearing capacity. The need to quantify these affects and the dynamic nature of structure must be done to develop more realistic predictive models.

The need for addressing the dynamic nature of soil structure in models has never been greater. We as soil scientists are being asked to predict environmental effects of many practices ranging from site-scale sediment transport and nutrient inputs from on-site wastewater to field-scale nutrient movement from agronomic, forestry, and development practices to landscape and watershed scale inputs from mixed land uses. Many models have used soil structure as a component but not as a dynamic component. This has not always resulted in successful predictions. The incorporation of soil structure as a dynamic component in new system models will overcome these limitations resulting in more accurate predictions and thus improved land management.

This publication is the outgrowth of a symposium entitled "Quantifying and Modeling Soil Structure Dynamics" held in 2009 at the SSSA Annual Meeting. Several authors and new cutting edge information have been added to the papers in this publication, making it the definitive work on structure dynamics and modeling to date.

David L. Lindbo, 2013 SSSA President

Soil structure is the arrangement and interconnectedness of soil aggregates/peds and of the pores. Soil structure is affected by tillage, wheel traffic, root growth, activity of microbes and mesofauna, shrink–swell, and freeze–thaw. Soil structure has an impact on root growth and function, soil micro- and mesofauna, solute transport, water infiltration, gas exchange, thermal and electrical conductivities, traffic bearing capacity, and more. Ignoring soil structure can lead to incorrect model predictions of runoff, erosion, solute leaching, root function, soil aeration, and groundwater recharge. Soil structure is now often included in these process-oriented models. Targeted simulation models have been developed to describe certain aspects of soil structure, such as network of pores, preferential flow, shrink–swell, and fractal analysis of soil aggregation. This book brings together these components to address systems in which correct description of soil structure is vital.

Soil structure should be included in system models such as solute leaching, erosion, and crop growth. Ignoring soil structure in the past resulted in models being unable to predict some aspects of soil behavior, such as the solute leaching that was observed and the functioning of roots clustered around soil peds. In the last 25 years, soil structure components have been recognized as vital to these processes, and incorporated into many simulation models. Nevertheless, soil structure has not been as well incorporated into larger scale items, such as pedotransfer function description of the soil water retention curve/hydraulic conductivity or watershed-scale water quality issues. Even when soil structure is included in a simulation model, the structure is often considered static rather than dynamic. The objectives of this book are to show the dynamic aspects of soil structure and how they can be included in system models.

Ideas for this volume came from a symposium entitled "Quantifying and Modeling Soil Structure Dynamics" at the 2009 International Annual Meetings of the American Society of Agronomy, Crop Science Society of America, and Soil Science Society of America. Some of those presentations had already been published and are not included here; additional authors and chapters have been added that were not part of the symposium.

Sally Logsdon, Markus Berli, and Rainer Horn, Editors

Contributors

Benjamin, Joseph G. Central Great Plains Research Station, Northern Plains Area, 40335 Rd. GG, Akron, CO 80720 (joseph.benjamin@ars.usda.gov)

Berli, Markus Desert Research Inst., 755 E. Flamingo Road, Las Vegas, NV 89119 (markus.berli@dri.edu)

Blanco-Canqui, Humberto Dep. of Agronomy and Horticulture, University of Nebraska, 261 Plant Science Hall, Lincoln, NE 68583-0915 (hblancocanqui2@unl.edu)

Dexter, Anthony R. Institute of Soil Science and Plant Cultivation (IUNG), ul. Czartoryskich 8, 24-100, Pulawy, Poland (tdexter@iung.pulawy.pl)

Fischer, Gottfried RIF e.V. Institut für Forschung und Transfer, Joseph-von-Fraunhofer-Str. 20, D-44227 Dortmund, Germany (gottfried.fischer@rif-ev.de)

Hallett, Paul D. The James Hutton Institute, Invergowrie, Dundee, DD2 5DA, United Kingdom (paul.hallett@hutton.ac.uk)

Horn, Rainier University of Kiel, Kiel University, Inst. of Soil Science, Plants, Nutrition, Olshausenstr. 40, Kiel, 24118, Germany (rhorn@soils.uni-kiel.de)

Korošak, Dean Applied Physics, Faculty of Civil Engineering, Univ. of Maribor, Smetanova ulica 17, Maribor SI-2000, Slovenia, and Faculty of Medicine, Inst. of Physiology, Univ. of Maribor, Slomškov trg 15, Maribor SI-2000, Slovenia (dean.korosak@uni-mb.si)

Levy, G.J. Inst. of Soil, Water and Environmental Sciences, Agricultural Research Organization, POB 6, Bet Dagan 50250, Israel (vwguy@volcani.agri.gov.il)

Logsdon, Sally USDA-ARS National Laboratory for Agriculture and the Environment, USDA-ARS National Lab. for Agric. and the Environ., 2110 University Boulevard, Ames, IA 50011 (sally.logsdon@ars.usda.gov)

Mamedov, A.I. USDA-ARS, Engineering and Wind Erosion Research Unit, 1515 College Ave., Kansas State Univ., Manhattan, KS 66502, USA (am03@ksu.edu)

Mooney, Sacha Jon School of Biosciences, Univ. of Nottingham, The Gateway Building, Sutton Bonington Campus, Leicestershire, LE12 5RD, UK (sacha.mooney@nottingham.ac.uk)

Nellesen, Jens RIF e.V. Institut für Forschung und Transfer, Joseph-von-Fraunhofer-Str. 20, D-44227 Dortmund, Germany (jens.nellesen@rif-ev.de)

Peth, Stephan Universität Kassel, Fachgebiet Bodenkunde, Nordbahnhofstraße 1a, D-37213 Witzenhausen, Germany (peth@uni-kassel.de)

Poulsen, Tjalfe G. Dep. of Chemistry and Biotechnology, Aalborg Univ., Sohngaardsholmsvej 57, DK-9000 Aalborg, Denmark (tgp@bio.aau.dk)

Tillmann, Wolfgang Technische Universität Dortmund, Faculty of Mechanical Engineering, Institute for Materials Engineering, Leonhard-Euler-Str. 2, D-44227 Dortmund, Germany (wolfgang.tillmann@tu-dortmund.de)

Yoshida, Shuichiro Dep. of Biological and Environmental Engineering, Graduate school of Agronomy and Life Science, Univ. of Tokyo, 1-1-1, Yayoi, Bunkyo, Tokyo, 113-8657, Japan (agyoshi@mail.ecc.u-tokyo.ac.jp)

Chapter 1

Quantifying and Modeling Soil Structure Dynamics

Sally Logsdon, Markus Berli, and Rainer Horn

What is soil structure, and how should it be characterized? The first part of this question was answered by Baver (1940) as simply "the arrangement of soil particles," a definition which, with minor modifications, still holds today (Ghezzehei, 2012). The second part of this question is more difficult to answer and has been a topic of scientific discussions ever since soil structure has been recognized as an important factor affecting soil physical, mechanical, chemical, and biological processes (Letey, 1991). Morphologic characterization of soil structure has been common practice in agronomy, forestry, and geomorphology as reflected in the soil structure classification systems (i.e., Schoeneberger et al., 2002). Beyond semi-quantitative soil morphology classes, it is a challenge to describe and quantify soil structure, per se. As pointed out by Letey (1991), the study of soil structure may be as well an art as a science.

Numerous papers compare management and experimental procedure effects on mechanical structured soil as well as single aggregate strength, intra-aggregate pore size distribution, and gas fluxes (Horn et al., 1994) as well as water stable aggregation (Bronick and Lal, 2005). The structure of a disturbed soil may be quite different than the structure in situ (Hartge and Stewart, 1995; Warkentin, 2008). Often the laboratory procedures to determine aggregate stability remove the aggregates from their in situ arrangement (Letey, 1991; Young et al., 2001; Baveye, 2006) and ignore the effect of aggregate arrangement and internal aggregate strength in comparison with that of the bulk soil (Horn et al., 1994).

Abbreviations: CT, computed tomography; NMR, nuclear magnetic resonance imaging; NR, neutron radiography; ptf, pedotransfer function; XMT, X-ray microtomography.

S. Logsdon, USDA-ARS National Laboratory for Agriculture and the Environment, USDA-ARS National Lab. for Agric. and the Environ., 2110 University Boulevard, Ames, IA 50011 (sally.logsdon@ars.usda.gov); M. Berli, Desert Research Institute, 755 E. Flamingo Road, Las Vegas, NV 89119 (markus.berli@dri.edu); R. Horn, University of Kiel, Kiel University, Institute of Soil Science, Plants, Nutrition, Olshausenstr. 40, Kiel, 24118, Germany (rhorn@soils.uni-kiel.de).

doi:10.2134/advagricsystmodel3.c1

Furthermore, only the comparison between the in situ bulk soil strength and that of single aggregates reveal the effects of wetting and drying, enhanced particle bonding and "gluing" mechanisms, and the consequences for a more rigid pore system and pore functioning (Horn and Baumgartl, 2002). Wet aggregate stability and wet and dry aggregate size distributions are often used as soil indicators related to infiltration as well as water and wind erosion. Sampled aggregates without a defined hydraulic history in relation to the in situ conditions give an incomplete picture of soil structure in relation to preferential flow, solute transport, gas exchange, and root growth patterns.

Some of the difficulties in direct-quantification of soil structure directly can be circumvented by quantifying the "interior" architecture (voids, biopores, and cracks between the aggregates and peds) rather than the solid "framework" of the soil (Letey, 1991; Warkentin, 2008). Rates of water movement, gas exchange, or solute transport through undisturbed soil represent indirect measures of soil structure (Edwards et al., 1993; Alaoui et al., 2011). Pedotransfer functions (ptfs) are used to relate soil properties to soil functions (such as water movement). The ptfs are more useful if morphologically defined structure classes as well as texture and bulk density are related to soil hydraulic properties (Espino et al., 1996; Vereecken et al., 2010).

One goal for quantifying soil structure is to predict hydraulic properties. The difficulty is determining what aspects of soil structure (size, shape, angularity) would be useful for predicting soil hydraulic properties. Based on measured stress-strain relationships and stress-induced changes in hydraulic properties, Horn and Fleige (2003) developed a set of ptfs which empirically connect morphologically determined structure classes with soil mechanical and hydraulic properties. Lin et al. (1999) used structural (peds and macropore characteristics) and physical properties (texture and density) in separate principal component analyses, and related them to flow in the macropore, mesopore, and micropore regions. The structural properties used to predict macropore flow, whereas the textural and density properties best predicted the micropore flow region. Water retention functions now often include two or more connected sections (macropore and matrix regions: Durner, 1994) to account for the effects of soil structure. Alaoui et al. (2011) reviewed how compaction and shrinkage affect the arrangement of pores and soil hydrology. They suggest further development is needed to upscale from the pore size processes to two region hydrology models.

Solute transport is strongly affected by soil structure characteristics. Gerke (2006) has thoroughly reviewed how solute transport models incorporate soil structure and preferential flow. An important research need is to develop and test procedures that independently determine the parameters needed for mac-ropore or fracture function in preferential flow models Allaire et al. (2009) summarized several procedures used to characterize preferential flow, which includes breakthrough curves, image analysis of dye tracing, particle and gas tracing, and hydraulic conductivity. Jarvis et al. (2009) developed a scheme to characterize the degree of macropore flow. Factors considered are texture, till-age, presence of biopores, perennial vegetation, manure addition, and subsoil compaction. How far the in situ textural properties coincide with the measured lab values is generally unknown, but flow and strength conditions in the field are altered to a great extent.

There are various other indices of soil structure. Some are based on morphol-ogy (Ball and Douglas, 2003; Kerry and Oliver, 2007). Others are based on aspects of the water retention curve (Dexter, 2004a, 2004b, 2004c; Reynolds et al., 2009). Yet other indices are based on aggregate stability factors (Six et al., 2000; Horn and Fleige, 2003; Nichols and Toro, 2011). Geeves et al. (1998) developed two indices based on predicted runoff.

One of the main obstacles toward quantifying structure dynamics was the limited capabilities to visualize soil structure without destroying the structure itself. In recent years, nondestructive imaging techniques such as X-ray computed tomography (CT) (Duliu, 1999; Ketcham and Carlson, 2001; Wildenschild et al., 2002; Cnudde et al., 2006; Taina et al., 2008; Lombi and Susini, 2009), nuclear mag-netic resonance imaging (NMR) (Hemminga and Buurman, 1997) and neutron radiography (NR) (Oswald et al., 2008) become available that allow to study soil structure and related physical processes nondestructively and non-invasively (Crestana and Vaz, 1998; Young et al., 2001). NMR and NR are particularly well suited to determine soil water distribution because they sense the presence of hydrogen, whereas CT is suitable for imaging materials based differences in X-ray attenuation of the constituents (e.g., soil solids and pores). CT has been widely used to image soil pores and pore networks in soil volumes at the multiple cen-timeter to decimeter scale. (See review by Taina et al. (2008), for details). X-ray Microtomography (XMT), essentially CT but with specimen at the millimeter to centimeter scale yielding images with spatial resolutions at the scale of a few micrometer (Macedo et al., 1999), has already been applied to soils for studying aggregate microstructure (Macedo et al., 1998; Peth et al., 2008; Sleutel et al., 2008; Menon et al., 2011), changes in inter-aggregate pores and contacts under com-pressive forces and their impact on hydraulic properties (Eggers et al., 2006, 2007; Carminati et al., 2007; Berli et al., 2008; Aravena et al., 2011), localized deformations under triaxial stress conditions (Viggiani et al., 2004) as well as fluid distribution, flow and transport processes (Clausnitzer and Hopmans, 2000; Culligan et al.,

2004; Al-Raoush and Willson, 2005a, 2005b; Wildenschild et al., 2005; Culligan et al., 2006; Brusseau et al., 2006, 2007; Carminati et al., 2007; Tippkotter et al., 2009; Menon et al., 2011) and soil-water interfaces (Culligan et al., 2004; Brusseau et al., 2006, 2007; Carminati et al., 2007). Considering the potential of nondestructive imaging by XMT and progress made regarding image processing and numerical modeling, XMT it is very likely to further improve our understanding about soil structure and its impact on mechanical and hydraulic properties.

Overview of Chapters

There are obstacles to overcome to quantify soil structure as a "state," which shows the challenge in quantifying the dynamics of soil structure. The aim of this book is to provide an overview on current approaches and technologies that shed light on soil structure dynamics and how to quantify it. Rather than merely describing soil structure, this book seeks to quantify soil structure in ways that can be incorporated into larger models. The book is a collection of articles addressing soil structure and the dynamics from various angles.

The first set of two chapters deal with overviews of soil structure characterization. The first article titled, "Impacts of soil organic carbon on soil physical behavior" discusses how soil organic matter affects soil structure, and processes that depend on soil structure (Chapter 2, Blanco-Canqui and Benjamin, 2013, this volume). The carbon in the soil influences soil structure by buffering excessive soil compaction, increasing aggregate strength, macropores, water repellency, and water holding capacity. Studies continue on the influence of crop residues on soil organic carbon and the specific pools of organic carbon that positively affect soil physical properties. The second article on "High energy moisture characteristics: linking between soil physical processes and structural stability" provides an excellent background in soil structure, how to characterize it, and how it is disrupted (Chapter 3, Mamedov and Levy, 2013, this volume). The chapter shows the relationship between aggregate stability and the distribution of soil pores, and shows when the soil surface can be considered as a stable rigid component of the system.

The next set of chapters concentrates on soil voids. "Using network models to describe the porous architecture of soils" gives the developing description of soil pore network models, and the relation to measured pore arrangements (Chapter 4, Korosak and Mooney, 2012, this volume). The complex pore network models are better at showing the long-range effect of soil pores as well as the linkages among them. They suggest the next step might be to relate the pore structure to microbiological function in soil pores and on pore surface. "A history of understanding crack propagation and the tensile strength of soil" provides an excellent introduction into fracture mechanics theory and its development to predict soil crack formation in field soils (Chapter 5, Hallett et al., 2013, this volume). The introduced fracture mechanics model shows great potential to describe crack formation and extension in natural soil. "Dynamics of soil macropore networks in response to hydraulic and mechanical stresses investigated by X-ray microtomography" describes the laboratory procedures to characterize pore arrangement, and subsequent data analysis (Chapter 6, Peth et al., 2013, this volume). To better understand internal stress-strain effects on fracture closure, they suggest that experiments should be conducted at multiple stress levels, for a range of initial and boundary conditions. Then the soil mechanics models should be linked with models that describe the function of a changing pore system.

Finally structure effect on soil function is described in "Gas permeability in soils as relation to soil structure and pore system characteristics" (Chapter 7, Poulsen, 2013, this volume). This chapter shows the relationship between gas permeability and gas-filled porosity in relation to the stability and connectivity of the pores.

Synthesis/Outlook

As mentioned above, other aspects of the functional relation of soil structure are discussed by Edwards et al. (1993), Gerke (2006), Alaoui et al. (2011), and Ghezzehei (2012), that is, water movement, gas exchange, or solute transport. Alaoui et al. (2011) suggested the greatest research needs were large-scale characterization of preferential flow, determining preferential flow for partially frozen soil, determining preferential flow of particles and gases and bacteria and non-aqueous liquids, using math tools to compare results from different techniques, nonintrusive field methods, better image analysis procedures, procedures to study flow inside a single macropore or network, better understanding of the interface between macropore and matrix, and standardization of procedures.

These chapters address future needs for soil structure research. Blanco-Canqui and Benjamin (Chapter 2, 2013, this volume) would like to see more research on specific pools of organic matter in relation to elasticity and "springiness" related to soil rebound from an applied load, how roots and hyphae networks enmesh soil minerals and reduce friction between aggregates, and how organic colloid charges influence bonding between organic and mineral colloids. Determining temporal and spatial changes of porosity between and within aggregates would help us achieve stable soil structure and associated favorable environmental impacts (Chapter 3, Mamedov and Levy, 2013, this volume). Korosak and Mooney (Chapter 4, 2013, this volume) ask how biological activity alters the local soil structure, biological mobility, microsite protection, and ecological balance. Hallett et al. (Chapter 5, 2013, this volume) suggest that further work on their elasto-plastic approach will link soil structure dynamics to water transport modeling. Peth et al. (Chapter 6, 2013, this volume) so far have tested one sample to related 3-D images of pores to the water retention curve, but they see the need to extend to replicates and a broad range of soils as well as relate to other hydraulic properties. Poulsen (Chapter 7, 2013, this volume) would like to see the relation between gas permeability, stability, and connectivity incorporated into models of soil structure. Advances in digital imaging, image analysis, and process-based modeling by finite element or discrete element methods will assist theory development that relates soil structure and pores to soil function.

References

Alaoui, A., J. Lipiec, and H.H. Gerke. 2011. A review of the changes in the soil pore systems due to soil deformation: A hydrodynamic perspective. Soil Tillage Res. 115–116:1–15. doi:10.1016/j. still.2011.06.002.

Allaire, S.E., S. Roulier, and A.J. Cessna. 2009. Quantifying preferential flow in soils: A review of different techniques. J. Hydrol. 378:179–204. doi:10.1016/j.jhydrol.2009.08.013.

Al-Raoush, R.I., and C.S. Willson. 2005a. A pore-scale investigation of a multiphase porous media system. J. Contam. Hydrol. 77:67–89. doi:10.1016/j.jconhyd.2004.12.001.

Al-Raoush, R.I., and C.S. Willson. 2005b. Extraction of physically realistic network properties from three-dimensional synchrotron Z-ray microtomography images of unconsolidated porous media systems. J. Hydrol. 300:44–64. doi:10.1016/j.jhydrol.2004.05.005.

Aravena, J.E., M. Berli, T.A. Ghezzehei, and S.W. Tyler. 2011. Effects of root-induced compaction on rhizosphehre hydraulic properties- X-ray micro-tomography imaging and numerical simulations. Environ. Sci. Technol. 45:425–431. doi:10.1021/es102566j.

Ball, B.C., and J.T. Douglas. 2003. A simple procedure for assessing soil structural, rooting and surface conditions. Soil Use Manage. 19:50–56. doi:10.1111/j.1475-2743.2003.tb00279.x.

Baver, L.D. 1940. Soil physics. John Wiley and Sons, New York.

Baveye, P. 2006. Comment on "Soil structure and management: A review" by C.J. Bronick and R. Lal. 2006. Geoderma 134:231–232. doi:10.1016/j.geoderma.2005.10.003.

Berli, M., A. Carminati, T.A. Ghezzehei, and D. Or. 2008. Evolution of unsaturated hydraulic conductivity of aggregated soils due to compressive forces. Water Resour. Res. 44:W00C09. doi:10.1029/2007WR006501.

Blanco-Canqui, H., and J.G. Benjamin. 2013. Impacts of soil organic carbon on soil physical behavior. In: S. Logsdon, M. Berli, and R. Horn, editors, Quantifying and modeling soil structure dynamics. Advances in Agricultural Systems Modeling 3. SSSA, Madison, WI. p. 11–40. doi:10.2134/advagricsystmodel3.c2

Bronick, C.J., and R. Lal. 2005. Soil structure and management: A review. Geoderma 124:3–22. doi:10.1016/j.geoderma.2004.03.005.

Brusseau, M.L., S. Peng, G. Schnaar, and M.S. Constanza-Robinson. 2006. Relationships among air-water interfacial area, capillary pressure, and water saturation for a sandy porous medium. Water Resour. Res. 42:W03501. doi:10.1029/2005WR004058.

Brusseau, M.L., S. Peng, G. Schnaar, and A. Murao. 2007. Measuring air-water interfacial areas with X-ray microtomography and interfacial partitioning tracer tests. Environ. Sci. Technol. 41:1956–1961. doi:10.1021/es061474m.

Carminati, A., A. Keastner, R. Hassanein, O. Ippisch, P. Vontobel, and F. Flühler. 2007. Infiltration through series of soil aggregates: Neutron radiography and modeling. Adv. Water Resour. 30:1168–1178. doi:10.1016/j.advwatres.2006.10.006.

Clausnitzer, V., and J.W. Hopmans. 2000. Pore-scale measurements of solute breakthrough using microfocus X-ray computed tomography. Water Resour. Res. 36:2067–2079. doi:10.1029/2000WR900076.

Cnudde, V., B. Masschaele, M. Dierick, J. Vlassenbroeck, L. Van Hoorebeke, and P. Jacobs. 2006. Recent progress in X-ray CT as a geosciences tool. Appl. Geochem. 21:826–832. doi:10.1016/j.apgeochem.2006.02.010.

Crestana, S., and C.M.P. Vaz. 1998. Non-invasive instrumentation opportunities for characterizing soil porous systems. Soil Tillage Res. 47:19–26. doi:10.1016/S0167-1987(98)00068-3.

Culligan, K.A., D. Wildenschild, B.S.B. Christensen, W.G. Gray, and M.L. Rivers. 2006. Pore-scale characteristics of multiphase flow in porous media: A comparison of air-water and oil-water experiments. Adv. Water Resour. 29:227–238. doi:10.1016/j.advwatres.2005.03.021.

Culligan, K.A., D. Wildenschild, B.S.B. Christensen, W.G. Gray, M.L. Rivers, and A.F.B. Tompson. 2004. Interfacial area measurements for unsaturated flow through a porous medium. Water Resour. Res. 40:W12413. doi:10.1029/2004WR003278.

Dexter, A.R. 2004a. Soil physical quality Part I. Theory, effects of soil texture, density, and organic matter, and effects on root growth. Geoderma 120:201–214. doi:10.1016/j.geoderma.2003.09.004.

Dexter, A.R. 2004b. Soil physical quality Part II. Friability, tillage, tilth, and hard-setting. Geoderma 120:215–225. doi:10.1016/j.geoderma.2003.09.005.

Dexter, A.R. 2004c. Soil physical quality Part III. Unsaturated hydraulic conductivity and general conclusions about S-theory. Geoderma 120:227–239. doi:10.1016/j.geoderma.2003.09.006.

Duliu, O.G. 1999. Computer axial tomography in geosciences: An overview. Earth Sci. Rev. 48:265–281. doi:10.1016/S0012-8252(99)00056-2.

Durner, W. 1994. Hydraulic conductivity estimation for soils with heterogeneous pore structure. Water Resour. Res. 30:211–223.

Edwards, W.M., M.J. Shipitalo, and L.B. Owens. 1993. Gas, water, and solute transport in soils containing macropores: A review of methodolgy. Geoderma 57:31–49. doi:10.1016/0016-7061(93)90146-C.

Eggers, C.G., M. Berli, M.L. Accorsi, and D. Or. 2006. Deformation and permeability of aggregated soft earth materials. J. Geophys. Res. 111:B10204. doi:10.1029/2005JB004123.

Eggers, C.G., M. Berli, M.L. Accorsi, and D. Or. 2007. Permeability of deformable soft aggregated earth materials: From single pore to sample cross-section. Water Resour. Res. 43:W08424. doi:10.1029/2005WR004649.

Espino, A., D. Mallants, M. Vanclooster, and J. Feyen. 1996. Cautionary notes on the use of pedo-transfer functions for estimating soil hydraulic properties. Agric. Water Manage. 29:235–253. doi:10.1016/0378-3774(95)01210-9.

Gerke, H.H. 2006. Preferential flow for structured soils. J. Plant Nutr. Soil Sci. 169:382–400. doi:10.1002/jpln.200521955.

Geeves, G.W., H.P. Cresswell, and B.W. Murphy. 1998. Two indices of soil structure based on prediction of soil water processes. Soil Sci. Soc. Am. J. 62:223–232. doi:10.2136/sssaj1998.03615995006200010029x.

Ghezzehei, T.A. 2012. Soil structure. In: P. Huang et al., editors, Handbook of soil science. CRC Press, Boca Raton, FL. p. 1–17.

Hallett, P.D., A.R. Dexter, and S. Yoshida. 2013. A history of understanding crack propagation and the tensile strength of soil. In: S.D. Logsdon, M. Berli, and R. Horn, editors, Quantifying and modeling soil structure dynamics. Advances in Agricultural Systems Modeling 3. SSSA, Madison, WI. p. 93–120. doi:10.2134/advagricsystmodel3.c5

Hartge, K.H., and B.A. Stewart, editors. 1995. Soil structure its development and function. Advances in soil science. CRC Lewis Pub., New York.

Hemminga, M.A., and P. Buurman. 1997. NMR in soil science. Geoderma 80:221–224. doi:10.1016/S0016-7061(97)00053-0.

Horn, R., and T. Baumgartl. 2002. Dynamic properties of soils. In: A.W. Warrick, editor, Soil physics companion. CRC Press, Boca Raton, FL. p. 17–48.

Horn, R., and H. Fleige. 2003. A method of assessing the impact of load on mechanical stability and on physical properties of soils. Soil Tillage Res. 73:89–100. doi:10.1016/S0167-1987(03)00102-8.

Horn, R., H. Taubner, M. Wuttke, and T. Baumgartl. 1994. Soil physical properties related to soil structure. Soil Tillage Res. 30:187–216. doi:10.1016/0167-1987(94)90005-1.

Jarvis, N.J., J. Moeys, J.M. Hollis, S. Reichenberger, A.M.L. Lindahl, and I.G. Dubus. 2009. A conceptual model of soil susceptibility to macropore flow. Vadose Zone J. 8:902–910. doi:10.2136/vzj2008.0137.

Kerry, R., and M.A. Oliver. 2007. The analysis of ranked observations of soil structure using indicator geostatistics. Geoderma 140:397–416. doi:10.1016/j.geoderma.2007.04.020.

Ketcham, R.A., and W.D. Carlson. 2001. Acquisition, optimization and interpretation of X-ray computer tomographic imagery: Applications to the geosciences. Comput. Geosci. 27:381–400. doi:10.1016/S0098-3004(00)00116-3.

Korošak, D., and S.J. Mooney. 2013. Applications of complex network models to soil porous systems. In: S.D. Logsdon, M. Berli, and R. Horn, editors, Quantifying and modeling soil structure dynamics. Advances in Agricultural Systems Modeling 3. SSSA, Madison, WI. p. 75–92. doi:10.2134/advagricsystmodel3.c4

Letey, J. 1991. The study of soil structure: Science or art. Aust. J. Soil Res. 29:699–707. doi:10.1071/SR9910699.

Lin, H.S., K.J. McInnes, L.P. Wilding, and C.T. Hallmark. 1999. Effects of soil morphology on hydraulic properties: II. Hydraulic pedotransfer function. Soil Sci. Soc. Am. J. 63:955–960. doi:10.2136/sssaj1999.634955x.

Lombi, E., and J. Susini. 2009. Synchroton-based techniques for plant and soil science: Opportunities, challenges and future perspectives. Plant Soil 320:1–35. doi:10.1007/s11104-008-9876-x.

Macedo, A., S. Crestana, and C.M.P. Vaz. 1998. X-ray microtomography to investigate thin layers of soil clod. Soil Tillage Res. 49:249–253. doi:10.1016/S0167-1987(98)00180-9.

Macedo, A., C.M.P. Vaz, J.M. Naime, P.E. Cruvinel, and S. Crestana. 1999. X-ray microtomography to characterize the physical properties of soil and particulate systems. Powder Technol. 101:178–182. doi:10.1016/S0032-5910(98)00170-3.

Mamedov, A.I., and G.J. Levy. 2013. High energy moisture characteristics: Linking between some soil physical processes and structure stability. In: S.D. Logsdon, M. Berli, and R. Horn, editors, Quantifying and modeling soil structure dynamics. Advances in Agricultural Systems Modeling 3. SSSA, Madison, WI. p. 41–74. doi:10.2134/advagricsystmodel3.c3

Menon, M., Q. Yuan, Z. Jia, A.J. Dougill, S.R. Hoon, A.D. Thomas, and R.A. Williams. 2011. Assessment of physical and hydrological properties of biological soil crust using x-ray microtomography and modeling. J. Hydrol. 397:47–54. doi:10.1016/j.jhydrol.2010.11.021.

Nichols, K.A., and M. Toro. 2011. A whole soil stability index (WSSI) for evaluating soil aggregation. Soil Tillage Res. 111:99–104. doi:10.1016/j.still.2010.08.014.

Oswald, S.E., M. Menon, A. Carminati, P. Vontobel, E. Lehmann, and R. Schulin. 2008. Quantitative imaging of infiltration, root growth, and root water uptake via neutron radiography. Vadose Zone J. 7:1035–1047. doi:10.2136/vzj2007.0156.

Peth, S., R. Horn, F. Beckmann, T. Donath, J. Fisher, and A.J.M. Smucker. 2008. Three-dimensional quantification of intra-aggregate pore-space features using synchrotron radiation-based microtomography. Soil Sci. Soc. Am. J. 72:897–907. doi:10.2136/sssaj2007.0130.

Peth, S., J. Nellesen, G. Fischer, W. Tillman, and R. Horn. 2013. Dynamics of soil macropore networks in response to hydraulic and mechanical stresses investigated by X-ray microtomography. In: S.D. Logsdon, M. Berli, and R. Horn, editors, Quantifying and modeling soil structure dynamics. Advances in Agricultural Systems Modeling 3. SSSA, Madison, WI. p. 121–154. doi:10.2134/advagricsystmodel3.c6

Poulsen, T. 2013. Gas permeability in soils as related to soil structure and pore system characteristics. In: S.D. Logsdon, M. Berli, and R. Horn, editors, Quantifying and modeling soil structure dynamics. Advances in Agricultural Systems Modeling 3. SSSA, Madison, WI. p. 155–186. doi:10.2134/advagricsystmodel3.c7

Reynolds, W.D., C.F. Drury, C.S. Tan, C.A. Fox, and X.M. Yang. 2009. Use of indicators and pore volume-function characteristics to quantify soil physical quality. Geoderma 152:252–263. doi:10.1016/j.geoderma.2009.06.009.

Schoeneberger, P.J., D.A. Wysocki, E.C. Benham, and W.D. Broderson. 2002. Profile/pedon description: (Soil) structure. In: Field book for describing and sampling soils. Version 2.0. USDA-NRCS-NSSC, Lincoln, NE. p. 41–55.

Six, J., E.T. Elliott, and K. Paustian. 2000. Soil structure and soil organic matter: II. A normalized stability index and the effect of mineralogy. Soil Sci. Soc. Am. J. 64:1042–1049. doi:10.2136/sssaj2000.6431042x.

Sleutel, S., V. Cnudde, B. Masschaele, J. Vlassenbroek, M. Dierick, L. Van Hoorebeke, P. Jacobs, and S. De Neve. 2008. Comparison of different nano- and micro-focus X-ray computed tomography set-ups for the visualization of the soil microstructure and soil organic matter. Comput. Geosci. 34:931–938. doi:10.1016/j.cageo.2007.10.006.

Taina, I.A., R.J. Heck, and T.R. Elliot. 2008. Applications of X-ray computed tomography to soil science: A literature review. Can. J. Soil Sci. 88:1–20. doi:10.4141/CJSS06027.

Tippkotter, R., T. Eickhorst, H. Taubner, B. Gredner, and G. Rademaker. 2009. Detection of soil water in macropores of undisturbed soil using microfocus X-ray tube computerized tomography (µCT). Soil Tillage Res. 105:12–20. doi:10.1016/j.still.2009.05.001.

Vereecken, H., M. Weynants, M. Javaux, Y. Pachepsky, M.G. Schaap, and M.Th. van Genuchten. 2010. Using pedotransfer functions to estimate the van Genuchten-Mualem soil hydraulic properties: A review. Vadose Zone J. 9:795–820. doi:10.2136/vzj2010.0045.

Viggiani, G., N. Lenoir, P. Besuelle, M. Di Michiel, S. Marello, J. Desrues, and M. Kretzschmer. 2004. X-ray microtomography for studying localized deformation in fine-grained geomaterials under triaxila compression. C.R. Mec. 332:819–826.

Warkentin, B.P. 2008. Soil structure: A history from tilth to habitat. Adv. Agron. 97:239–272.

Wildenschild, D., J.W. Hopmans, M.L. Rivers, and A.J.R. Kent. 2005. Quantitative analysis of flow processes in a sand using synchrotron-based X-ray microtomography. Vadose Zone J. 4:112–126. doi:10.2113/4.1.112.

Wildenschild, D., J.W. Hopmans, C.M.P. Vaz, M.L. Rivers, and D. Rikard. 2002. Using X-ray computed tomography in hydrology: Systems, resolutions, and limitation. J. Hydrol. (Amersterdam) 267(3–4):285–297.

Young, I.M., J.W. Crawford, and C. Rappoldt. 2001. New methods and models for characterising structural heterogeneity of soil. Soil Tillage Res. 61:33–45. doi:10.1016/S0167-1987(01)00188-X.

Chapter 2

Impacts of Soil Organic Carbon
on Soil Physical Behavior

Humberto Blanco-Canqui and Joseph G. Benjamin

Abstract

Management-induced changes in soil organic carbon (SOC) concentration can affect soil physical behavior. Specifically, removal of crop residues as biofuel may thus adversely affect soil attributes by reducing SOC concentration as crop residues are the main source of SOC. Implications of crop residue management for soil erosion control, water conservation, nutrient cycling, and global C cycle have been discussed, but the potential impacts of residue removal-induced depletion of SOC on soil physical properties have not been widely studied. We reviewed published information on the relationships of SOC concentration with soil structural stability, consistency, compaction, soil water repellency, and hydraulic properties with emphasis on crop residue management. Our review indicates that studies specifically assessing relationships between crop residue management-induced changes in SOC concentration and soil physical properties are few. These studies indicate, however, that crop removal or addition can alter SOC concentration and concomitantly affect soil physical attributes with a magnitude depending on the amount of residue removed or returned, constituents of residue-derived SOC, tillage and cropping system, soil type, and climate. Our review also indicates that, in general, management practices that affect SOC concentration can directly influence soil physical properties. Decrease in SOC concentration reduces subcritical water repellency and aggregate stability and strength, increases soil's susceptibility to excessive compaction, and reduces macroporosity, hydraulic conductivity, and water retention. Soil organic matter improves soil physical properties by providing organic binding agents, inducing slight water repellency, lowering soil bulk density, and improving the elasticity and resilience of the whole soil. The numerous benefits of SOC on soil physical attributes suggest that SOC concentration should be maintained or increased through proper management practices. Indiscriminate residue removal for off-farm uses reduces SOC pools and can adversely affect soil and environment. Crop residues not only protect the soil surface from erosive forces but also maintain SOC concentration, which is essential to improve soil physical behavior and sustain soil productivity. Management practices including no-till with high residue input, continuous cropping systems, cover crops, and grass-based rotations should be promoted to further increase SOC concentration and thus improve soil physical behavior.

Abbreviations: SOC, soil organic carbon.

H. Blanco-Canqui, Department of Agronomy and Horticulture, University of Nebraska, 261 Plant Science Hall, Lincoln, NE 68583-0915 (hblancocanqui2@unl.edu); J.G. Benjamin, Central Great Plains Research Station, Northern Plains Area, 40335 Rd. GG, Akron, CO 80720 (joseph.benjamin@ars.usda.gov).

doi:10.2134/advagricsystmodel3.c2

SOIL ORGANIC CARBON

Induces Slight Water Repellency to Soils

Increases Soil Aggregate Stability and Strength

Increases Soil Consistency or Atterberg Limits

Reduces Bulk Density and Particle Density

Increases Total Porosity, Macroporosity and Pore Continuity

Increases Water Infiltration and Hydraulic Conductivity

Increases Soil Water Adsorption and Retention

M anagement practices including tillage, cropping systems, and crop residue removal or addition alter the concentration of organic C in the soil. The changes in SOC concentration may concomitantly impact soil physical attributes and soil productivity. Soil organic particles interact with inorganic particles to promote soil aggregation, increase porosity, and stabilize soil structure (Kay, 1997). Influence of soil organic matter on soil structure, nutrient cycling, C cycling, soil biological processes, and other ecosystem services has been studied (Weil and Magdoff, 2004), yet the mechanisms involved and the magnitude at which management-induced changes in SOC influence soil physical properties deserve further discussion.

Specifically, crop residue removal or addition dictates C input and SOC dynamics. At present, crop residues are confronted by a number of competing on- and off-farm uses. On one hand, crop residues are needed to conserve soil and water, reduce water and wind erosion, and maintain SOC concentration (Wilhelm et al., 2004). On the other hand, residues have potential off-farm uses including cellulosic ethanol production (Perlack et al., 2005), fiber production (Reddy and Yang, 2005), and livestock feed (Tanaka et al., 2005).

Influence of residue-management-induced SOC gains or losses on soil physical behavior such as structural stability, compactibility, and soil-water relationships has not been widely documented. Changes in soil physical properties and SOC concentration in residue management studies have often been discussed as static or separate parameters with little emphasis on the mutual interrelationships between soil structure and SOC. A synthesis of information on SOC vs. soil physical behavior relationships is needed to better understand the implications that crop residue management may have on soil physical properties. Correlations between soil structural properties and SOC concentration have been reported, but information is fragmented and has not been presented in a common framework applied to crop residue management.

Therefore, the specific objective of this chapter is to discuss the relationships of SOC with soil structural stability, consistency, compaction, soil water repellency, and hydraulic properties based on published studies with emphasis on crop residue management. We reviewed (i) published studies, which assessed the independent effects of crop residue management on soil physical properties and SOC concentration and (ii) relevant studies reporting information on SOC vs. soil structure correlations under integrated tillage-crop-residue management systems. This Chapter discusses factors affecting SOC vs. soil structure relationships and the specific relationships of SOC concentration with soil physical properties. Because impacts of crop residue removal on changes in SOC pools have been previously reviewed (Blanco-Canqui and Lal, 2008a; Anderson-Teixeira et al., 2009), we have only briefly discussed this topic.

Crop Residues and Soil Organic Carbon

Crop residues are the main source of soil organic C. On a dry-weight basis, crop residues contain, on average, 45% C (Lal, 1997; Blanco-Canqui and Lal, 2008a). Residue removal for off-farm uses reduces the amount of residue returned to soil, and it can thus alter C recycling and rapidly reduce SOC concentration. Residue removal directly reduces the SOC concentration by reducing C input. Recent reviews on the potential impacts of crop residue removal on SOC storage have indicated that excessive removal of residues may reduce SOC concentration (Blanco-Canqui and Lal, 2008a; Anderson-Teixeira et al., 2009). Residue removal may reduce SOC concentration even within a few years after removal (Blanco-Canqui and Lal, 2008a). The magnitude of SOC loss with residue removal may vary with soil type, tillage and cropping system, and climate; but, in the long term, excessive residue removal can result in a net SOC loss (Blanco-Canqui and Lal, 2008a).

Return or addition of crop residues maintains and increases SOC concentration. In soils where crop residues are removed for off-farm uses, adoption of intensive cropping systems with inclusion of cover crops and high biomass producing crops under no-till farming can provide additional residue input and restore the SOC lost with residue removal. Recent studies have shown that no-till continuous cropping systems provide more residue input and have greater SOC concentration than crop-fallow and crop-crop-fallow systems (Benjamin et al., 2008; Blanco-Canqui et al., 2010a). Similarly, addition of cover crops to no-till systems can increase SOC concentration with a magnitude depending on cover crop type, tillage management, and climate (Liebig et al., 2002; Liu et al., 2005; Villamil et al., 2006).

The above potential changes in SOC concentration under different scenarios (removal and addition) of residue management may concomitantly affect soil physical behavior including compaction, structural, and hydraulic parameters. Indeed, a key determinant for the degradation of soil physical properties following crop residue removal may be the loss of SOC. Factors that affect the magnitude of influence of residue removal-induced SOC loss on soil physical properties deserve discussion to better understand interactions and soil-specific response to crop residue management.

Factors that Affect Relationships Between Organic Carbon and Soil Physical Properties

The extent at which changes in SOC concentration affect soil physical properties depends on various interacting factors including climatic conditions, amount and constituents of SOC, textural class, tillage management, and others (Fig. 2–1). For example, climate in interaction with tillage and cropping systems directly influences crop residue production and rates of soil organic matter decomposition (Benjamin et al., 2008). The numerous interacting factors make the characterization of SOC influence on soil physical and hydraulic properties somewhat difficult.

Amount and Constituents of Soil Organic Carbon

Both amount and form of SOC influence soil physical behavior. A narrow range of SOC concentrations among residue management systems may have reduced or no effect on soil physical properties. In the central Great Plains, correlation

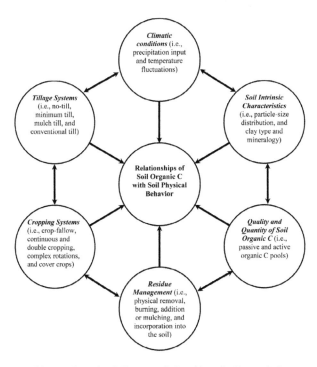

Fig. 2–1. Factors and interactions that influence relationships of soil organic C concentration with soil structural, compaction, consistency, mechanical, and hydraulic properties.

between SOC concentration and other soil physical properties was weak across permanent grass, wheat (*Triticum aestivum* L.), corn (*Zea mays* L.), millet (*Panicum liliaceum* L.), and wheat-fallow systems under no-till in spite of large differences in residue input after 15 yr of management (Benjamin et al., 2008). In some soils, however, even small changes in SOC can affect soil physical properties (Boivin et al., 2009).

It is also important to consider the different constituents of SOC and their role in influencing soil physical behavior. Only some constituents of total SOC may influence soil physical properties. Thus, total SOC may not be always strongly correlated with soil physical properties (Benjamin et al., 2008). Labile SOC or particulate organic matter may influence macroaggregation more than total SOC (Bell et al., 1998; Haynes, 2005; Dexter et al., 2008). Labile SOC can also be more sensitive to residue management than total SOC. Soil aggregates near the soil surface (e.g., no-till) are often stabilized by labile SOC fractions, while those at deeper soil horizons may be stabilized by persistent or recalcitrant SOC fractions with low turnover rates. Dexter et al. (2008) suggested "complexed organic C," which is the relation between 1 g of organic C and *n* g of clay, is the fraction that dominates the SOC effects on soil matrix physical properties relative to the amount of total SOC. More effort needs to be concentrated on the identification of specific fractions of SOC that affect soil physical properties under different scenarios of crop residue management.

Textural Class

Correlation between SOC concentration and soil physical parameters can be soil specific (Tables 2–1 and 2–2). Soil particle-size distribution is one of the major factors that impacts the extent at which SOC affects soil physical properties. For example, an increase in clay content may alter or diminish the beneficial effects of SOC on soil physical behavior (Kay, 1997; Saxton and Rawls, 2006). The SOC in interaction with clay particles forms organo-mineral combinations and creates aggregates (Malamoud et al., 2009). Table 2–2 shows that decreased SOC concentration with stover removal increased bulk density in silt loams but not in a clay loam in the 0- to 5-cm depth. Differences in particle-size distribution also influence SOC impacts on hydraulic properties. Table 2–2 shows that a reduction of SOC concentration resulted in lower saturated hydraulic conductivity in silt loams but not in a flat clay loam. Clayey soils normally have lower saturated hydraulic conductivity than medium- or coarse-textured soils. Soil hydraulic properties in clayey soils can be less sensitive to small changes in SOC concentrations than in sandy soils, but large increases in SOC concentration can affect soil hydraulic properties regardless of soil texture (Rawls et al., 2003; Rawls et al., 2004).

Table 2-1. Some examples of relationships of SOC with soil physical properties (≤20 cm depth).

Reference	Soil	Country	Management	Predictive equation	r^2
Aggregate stability					
Soane (1990) and Chaney and Swift (1984)	26 soils	Great Britain	Agricultural use	MWD (mm×100) = 26.8 + 23.2*SOM (%)	0.79***
Blanco-Canqui and Lal (2007)	Silt loam	United States	Straw management in no-till	MWD (mm) = -3.29 + 1.24*SOC (g kg⁻¹)	0.96***
Pikul et al. (2009)	Silty clay loam	United States	No-till and plow till	WSA (%) = -8.70 + 2.47*FPOM/SOM	0.64***
Jordán et al. (2010)	Loam	Spain	Straw management in no-till	AG (number of raindrops) = 13215 + 37729*SOM (%) + 08566(SOM)²	0.91***
Bulk density and proctor maximum bulk density					
Soane (1990) and Ball et al. (1989)	Two loams	Great Britain	No-till and plow till	Proctor ρ_{bmax} (Mg m⁻³) = 2.07-0.102*SOC (%)	0.93***
Quiroga et al. (1999)	24 sites (sand, loamy sand, sandy loam, and loam)	Argentina	Conventional till	Proctor ρ_{bmax} (Mg m⁻³) = 1.75–0.01*SOM (g kg⁻¹)	0.52***
Diaz-Zorita and Grosso (2000)	26 sites (loamy sand, loamy, and loamy silt)	Argentina	Conservation tillage and grasslands	Proctor ρ_{bmax} (Mg m⁻³) = 1.74-0.01*TOC (%)	0.75***
Blanco-Canqui and Lal (2007)	Silt loam	United States	Straw management in no-till	ρ_b (Mg m⁻³) = 4.39*(SOC, g kg⁻¹)⁻⁰·⁴⁹	0.87***
Particle density					
Blanco-Canqui et al. (2006b)	Silt loam	United States	No-till and plow till	ρ_s (Mg m⁻³) = 2.57- 0.004*SOC (g kg⁻¹)	0.62***
Blanco-Canqui and Lal (2007)	Silt loam	United States	Straw management in no-till	ρ_s (Mg m⁻³) = 2.87- 0.01*SOC (g kg⁻¹)	0.89***
Water repellency					
Blanco-Canqui and Lal (2007)	Silt loam	United States	Straw management in no-till	WDPT (s) = -2.15 + 0.09*SOC (g kg⁻¹)	0.52**
Blanco-Canqui (2011)	11 soils	United States	No-till and plow till	LogWDPT (s) = 0.31 + 0.01*SOC (g kg⁻¹)	0.18***
Soil consistency					
Eynard et al. (2006)	Loam and silt loam	United States	Plow till and grasslands	PL (kg kg⁻¹) = 189 + 2.5*SOC (g kg⁻¹)	0.89***
Blanco-Canqui et al. (2006b)	Silt loam	United States	No-till and plow till	LL (kg kg⁻¹) = 20.67 + 0.83*SOC (g kg⁻¹)	0.92***
				PL (kg kg⁻¹) = 14.06 + 0.58*SOC (g kg⁻¹)	0.88***

** Significant at the 0.01 probability level.
*** Significant at the 0.001 probability level.
† MWD, mean weight diameter of aggregates; AG, aggregate stability; WSA, water-stable aggregates; FPOM, fine particulate organic matter; SOM, soil organic matter; SOC, soil organic C; TOC, total organic C; ρ_s, particle density; ρ_b, bulk density; Proctor ρ_{bmax}, Proctor maximum bulk density; WDPT, water drop penetration time; PL, plastic limit; LL, liquid limit.

Table 2–2. Correlation coefficients of corn stover-derived soil organic C concentration with bulk density (ρ_b), log tensile strength of air-dry aggregates (Log TS), volumetric water content at different matric potentials, plant available water (PAW), log air permeability (Log k_a), and log saturated hydraulic conductivity (Log K_{sat}) after 1 yr of corn stover removal at 0, 25, 50, 75, and 100% removal in the 0- to 7.5-cm depth from three no-till soils in the eastern United States. Data from Blanco-Canqui et al. (2006a, 2007).

	ρ_b	Log TS	Volumetric water content			PAW	Log (k_a)	Log (K_{sat})
			−3 kPa	−30 kPa	−1,500 kPa			
	Mg m⁻³	kPa	—————— m³ m⁻³ ——————				μm²	mm h⁻¹
Silt loam (10% slope)	−0.75***	0.49*	0.76***	0.82**	0.68**	0.76***	0.62*	0.65**
Silt loam (<2% slope)	−0.41ns†	0.73**	0.73**	0.86***	0.86***	0.57*	0.72**	0.54*
Clay loam (<1% slope)	−0.52*	0.78***	0.65**	0.65**	0.65**	0.55*	0.67**	0.28ns

* Significant differences at 0.05 probability levels.
** Significant differences at 0.01 probability levels.
*** Significant differences at 0.001 probability levels.
† ns, not significant.

Tillage Management

Tillage impacts soil structure vs. SOC relationships by affecting the dynamics of crop residue decomposition and altering the proportion of organic binding agents (Table 2–1). Residues are left on the soil surface in no-till and minimum till and decompose slower than in conventional till where residues are plowed under and often rapidly decomposed. Consequently, SOC in no-till soils is often concentrated near the soil surface, while, in plowed soils, SOC is rather uniformly distributed in the plow layer. Benjamin et al. (2010) showed an increase in SOC in the surface 15 cm of soil of a no-till cropping system after 7 yr of continuous cropping under irrigation compared with a decrease in SOC in the surface 15 cm of a chisel plowed cropping system. In the 15- to 30-cm depth increment, the no-till system showed a decrease in SOC while the chisel plow system showed a SOC increase. The overall storage of SOC in the 0- to 30-cm soil depth, however, was greater with the no-till system. The amount of crop residues needed for maintaining SOC will vary according to tillage type and frequency and many crop species will not supply sufficient crop residues to improve SOC levels (Fig. 2–2).

Correlations between SOC and soil physical parameters in no-till soils are often stronger near the soil surface than in layers below. At deeper soil depths, increased clay content may dominate the changes in soil physical properties and minimize the impacts of SOC concentration. No-till continuous cropping systems with sod-based rotation or deep-rooted plant species, which increase SOC deeper in the soil profile can improve soil properties at or below the plow layer. On a

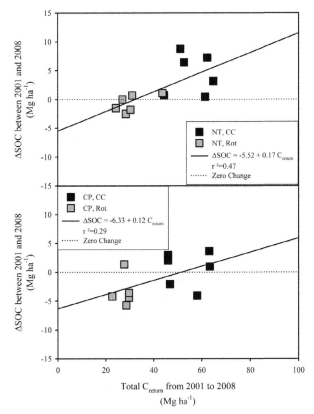

Fig. 2–2. Relationship of added crop residue C plus estimated added root and rhizodeposition C (C_{return}) on changes in soil organic C (ΔSOC) in the 0- to 30-cm depth increment between 2001 and 2008. NT denotes the no-till cropping system. The CP denotes the chisel plow cropping system. CC denotes the continuous corn rotation. Rot denotes the mixed grass and broadleaf crop rotation. (From Benjamin et al., 2010.)

loam in the central Great Plains, there was no significant correlation between SOC concentration and macroaggregates in the 0- to 18-cm depth, but, in the 20- to 37-cm depth, macroaggregates were positively correlated, although weakly, with differences in SOC concentration across cropping systems with different amounts of annual biomass C input (Benjamin et al., 2008). Differences in root growth patterns and interactions between SOC and clay fractions may affect soil aggregation at deeper depths. Further assessment of SOC vs. soil structure relationships for the whole soil profile is needed to understand how different scenarios of crop residue management influence soil properties.

Tillage also affects the nature and partitioning of organic binding agents that affect soil aggregation and stability of aggregates. Plowing reduces the proportion of temporary and transient organic binding agents through a rapid oxidization of soil organic matter. Tillage operations may not, however, affect persistent binding agents, which are resistant to degradation. The persistent agents are found as aromatic and organo-mineral complexes (Tisdall and Oades, 1982).

Interactions Between Soil Organic Carbon and Soil Physical Properties

Soil organic C and soil physical properties are linked together. Figure 2–3 shows a general flow of dynamic interactions of SOC with soil structural, compaction, and hydraulic properties. First, accumulation of SOC forms and stabilizes aggregates by providing organic binding agents and by inducing slight hydrophobic properties to soil (Fig. 2–4). Recent studies have indicated that development of slight water repellency may be essential to aggregate stability because it can reduce rapid air-pressure build-up inside the aggregates, which causes aggregate slaking (Goebel et al., 2005; Bottinelli et al., 2010; Blanco-Canqui, 2011). Second, well-aggregated soils have lower bulk density, better consistency, more macropores, and lower risks of excessive compaction than soils with low aggregate stability. Third, increased aggregation and macroporosity leads to improved water flow and retention characteristics. There is a mutual relationship between SOC and soil physical characteristics. On one hand, an increase in SOC concen-

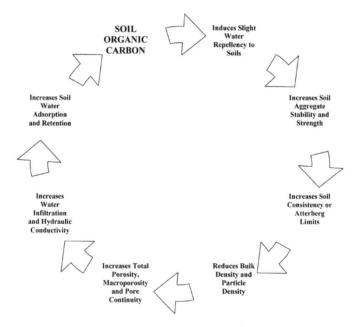

Fig. 2–3. A process diagram showing the linkages and interactions influencing relationships among soil organic C and soil physical properties.

tration promotes aggregation, reduces soil compactibility, and improves soil hydraulic properties. On the other hand, improved soil structural properties promote SOC protection and storage, which is essential to long-term C sequestration and overall soil productivity. The specific relationships of SOC with soil physical properties are discussed in the following sections.

Effect of Organic Carbon on Soil Water Repellency

Crop-residue derived SOC may induce some hydrophobic properties to soil (Table 2–1). While excessive soil water repellency can adversely affect soil structure and hydrology (Doerr et al., 2000; MacDonald and Huffman, 2004), slight water repellency observed in cultivated soils can have positive impacts on aggregate stabilization and long-term C sequestration (Hallett et al., 2001; Eynard et al., 2006; Lamparter et al., 2009). Residues are a food source for decomposers including earth-

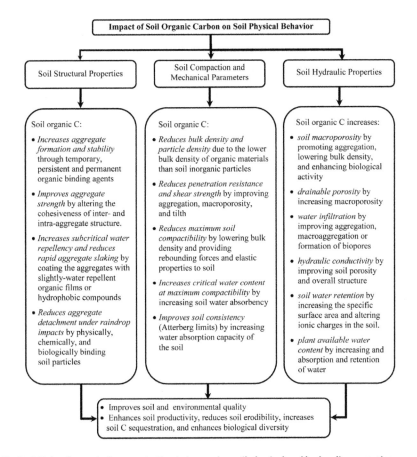

Fig. 2–4. Role of organic C concentration in improving soil physical and hydraulic properties.

worms, bacteria, fungi, and many others. Exudates from the activity of these macro and microorganisms coupled with decomposed organic materials (e.g., humic acids, aliphatic substances, and organic polymers) derived from crop residues can induce hydrophobic properties in soil aggregates (Chenu et al., 2000). Organic films coat soil aggregates and slow water entry into the aggregates, thereby reducing rapid aggregate slaking and detachment. Removal of residue reduces both SOC-enriched organic materials and microbial activity that induce hydrophobic properties to soil. On a plowed sandy soil after 22 yr of barley (*Hordeum vulgare* L.) straw removal, a decrease in SOC concentration by 1.2 times was associated with reduced soil water repellency (De Jonge et al., 2007). Barley straw contains complex waxes that induce water repellent properties to soil (De Jonge et al., 2007).

We used published data (Blanco-Canqui and Lal, 2008b) collected after 3 yr from three corn stover removal experiments established on a sloping silt loam, nearly level silt loam, and clay loam in eastern U.S. to further study the impacts of residue-removal-induced decrease in SOC concentration on soil water repellency. Correlations between changes in SOC concentration due to residue removal at different rates and air-dry aggregate water repellency as measured by water drop penetration time test were performed. Results from the correlations shown in Fig. 2–5A to 2–5C indicate that SOC concentration was moderately and positively correlated with water drop penetration time in the nearly level clay loam (Fig. 2–5C) but not in the silt loams (Fig. 2–5A and 2–5B). The significant correlation for the nearly level clay loam, but not for the silt loams suggests differences in clay concentration among the three soils probably influenced the SOC effect on water repellency. The range of variations in SOC concentration among the stover treatments within each soil was similar but the clay concentrations differed.

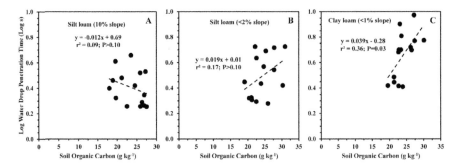

Fig. 2–5. Effect of soil organic C on soil water repellency as measured by the water drop penetration time test in a (A) sloping silt loam, (B) nearly level silt loam, and (C) flat clay loam in the 0- to 10-cm soil depth under corn stover removal at different rates from no-till soils in the eastern United States (data from Blanco-Canqui and Lal, 2009).

Interaction between clay content and SOC concentration most probably increases soil hydrophobicity in clayey soils over clay or SOC concentration alone. Association of SOC or humic compounds with clay minerals has been found to increase soil hydrophobicity in clayey soils (Chenu et al., 2000; Rodriguez-Alleres et al., 2007). Particularly, recalcitrant SOC fractions associate with the finest clay fractions and induce high hydrophobicity (Spaccini et al., 2002). Predominant factors that influence the manifestation of soil water repellency include SOC, clay concentration, and soil matric potential (De Jonge et al., 2007; Blanco-Canqui and Lal, 2008b).

Increase in SOC concentration with intensive cropping systems with high residue input also increases water repellency. Continuous cropping systems with conservation tillage, which leave residues on the soil surface, can accumulate SOC near the soil surface and induce water repellency to soils. In a 33-yr cropping system experiment in the central Great Plains, continuous wheat had 5 times greater aggregate water repellency than the average across sorghum-fallow, wheat-sorghum [Sorghum bicolor (L.) Moench]-fallow, continuous sorghum, and wheat-fallow under no-till for the 0- to 2.5-cm soil depth (Blanco-Canqui et al., 2010a). The hydrophobicity of residue-derived SOC varies with the quality of crop residues. De Jonge et al. (2007) observed that soils under barley and potatoes (Solanum tuberosum L.) had slightly greater soil water repellency than those under rye (Secale cereal L.), wheat, and corn. They also observed that grass plots had consistently greater soil water repellency than cropped systems at all soil water contents. Overall, changes in SOC concentration with residue removal or addition may change the hydrophobicity of soil, depending on the quantity and quality of residues. More experimental data on the impacts of crop residue management on soil water repellency are needed.

Effect of Organic Carbon on Soil Structural Stability And Strength

Influence of organic matter on soil aggregation has been widely discussed (Tisdall and Oades, 1982; Chaney and Swift, 1984; Weil and Magdoff, 2004; Fig. 2–4), but discussion on the specific impacts of crop residue removal or addition on SOC vs. soil structure relationships is somewhat limited (Table 2–1). Crop residues may differ on their impacts on soil structure from other amendments (e.g., animal manure, sawdust, and compost) as SOC influences on soil structure depend on the type and quality of amendments (Bhogal et al., 2009). Loss of SOC with residue removal may have a greater impact on soil structural parameters such as aggregate stability and strength than on other soil physical properties (Sparrow et al., 2006).

Aggregate stability decreases with a decrease in SOC concentration due to crop residue removal regardless of crop type. On two no-till silt loams in Iowa, percentage of wet aggregate stability decreased as the SOC concentration decreased in the 0- to 5-cm depth after 10 yr of stover management (Karlen et al., 1994). Across three no-till soils in the eastern U.S., decreases in SOC concentration following corn stover removal reduced mean weight diameter of aggregates in one of the three soils after 1 yr of removal (Fig. 2–6A to 2–6C), but, 3 yr after removal, it reduced mean weight diameter of aggregates in all soils in the 0- to 5-cm depth (Fig. 2–6D to 2–6F). Continued corn stover removal with time may reduce SOC concentration and thus aggregate stability regardless of soil type. On a plow-till clay loam in Canada, complete spring barley straw removal reduced diameter of water-stable aggregates from 2.3 to 1.3 mm and SOC concentration from 60.5 to 55.2 g kg^{-1} after 9 yr of straw management under conventional till (Singh et al., 1994). Macroaggregation decreases with crop residue-removal-induced decrease in SOC concentration. On a sandy loam in Texas, after 10 yr of

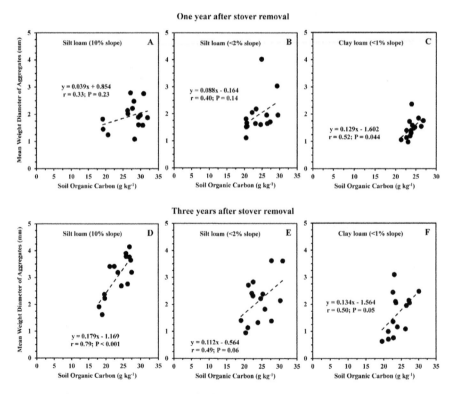

Fig. 2–6. Relationship between mean weight diameter of aggregates and soil organic C in the 0- to 5-cm soil depth for three no-till soils in the eastern United States after (A, B, and C) 1 yr and (D, E, and F) 3 yr of corn stover removal at different rates (data from Blanco-Canqui et al., 2006a; Blanco-Canqui and Lal, 2009).

residue management, wheat and sorghum residue removal from irrigated and dryland soils reduced both the proportion of macroaggregates from 29.7 to 25.3 g kg^{-1} and SOC concentration from 85 to 64.5 g kg^{-1} (Bordovsky et al., 1999).

Application of crop residues has the opposite effect to residue removal because it increases SOC concentration and it thus improves aggregate stability. On a silt loam in Ohio, increase in SOC concentration was linearly related (r = 0.50) to the increase in percentage of water-stable aggregates in the 0- to 10-cm depth after a 7-yr wheat straw application at five different rates to no-till, plow-till, and ridge-till soils (Duiker and Lal, 1999). On a loam in Spain, an increase in SOC concentration with the application of wheat straw at five different levels increased wet aggregate stability, explaining 91% of its variability in a no-till soil in the 0- to 10-cm depth in a 3-yr study (Jordán et al., 2010). Addition of by-products of corn stover fermentation can also increase soil aggregate stability by increasing SOC concentration. Johnson et al. (2004) reported that addition of stover fermentation by-product with 486 g kg^{-1} of SOC concentration linearly increased the aggregate stability, explaining 98% of its variability.

The main mechanisms by which crop residue removal reduces the stability of wet aggregates is by reducing the amount of organic binding agents and hydrophobicity of aggregates (Tisdall and Oades, 1982; Fig. 2–4). As discussed earlier, slight or subcritical water repellency can contribute to aggregate stabilization (Goebel et al., 2005; Bottinelli et al., 2010). Crop residues are a source of transient, temporary, and persistent organic binding agents that are essential to soil aggregation. Transient or labile soil organic matter fractions first bind soil particles into aggregates while the persistent or recalcitrant soil organic matter fraction, often occluded inside aggregates, contributes to permanent stabilization of soil structure (Kay, 1997; Weil and Magdoff, 2004). Constituents of soil organic matter, particularly persistent fractions, react with polyvalent cations, oxides, and aluminosilicates to form complex compounds and stabilize aggregates (Tisdall and Oades, 1982).

The SOC concentration and soil aggregates are mutually interrelated (Bossuyt et al., 2005). The SOC-enriched organic materials form and stabilize aggregates by providing organic binding agents, while aggregates in turn occlude and prevent SOC from rapid decomposition. Weaker aggregates store less SOC than more stable aggregates. Macroaggregate-protected SOC is mostly labile and young with faster turnover rates than micro-aggregate-protected SOC (Puget et al., 2005). Labile SOC fractions decrease more rapidly than stable or recalcitrant SOC fractions following residue removal (Karlen et al., 1994).

The degree at which the residue-derived SOC associates with soil mineral particles and stabilizes aggregates depends on the degree of residue decomposition (Kay, 1997). The association of residue-derived organic materials with soil mineral particles increases as crop residues decompose. Decomposed residues associate with microbial biomass and other organic compounds to glue and form aggregates. Presence of residue mulch on the soil surface is vital to provide a continuous supply of organic binding agents.

Residue-removal-induced loss of SOC not only decreases stability of wet aggregates but also reduces the strength and cohesiveness of dry aggregates. On a plow-till clay loam in Canada, complete spring barley straw removal reduced dry aggregate stability primarily by reducing SOC concentration after 9 yr of straw management under conventional till (Singh et al., 1994). In the eastern U.S., a decrease in SOC concentration with corn stover removal from no-till soils was reduced aggregate tensile strength in the 0- to 5-cm soil depth regardless of soil type within 1 yr after stover removal (Table 2–2). The decrease in dry aggregate strength with the decrease in SOC concentration is primarily attributed to the interactive effects among SOC, clay content, and oxides (Imhoff et al., 2002). Crop-residue derived organic materials and microbial exudates react with clay particles, fill the intra-aggregate pore spaces and, on drying, cement the contact points between organic and inorganic soil particles, increasing the strength of aggregates. The drying of aggregates associated with crop-residue derived organic materials increases the cohesion forces among particles (Imhoff et al., 2002).

Effect of Organic Carbon on Soil Consistency

Changes in SOC concentration may also affect soil consistency (Table 2–1). The Atterberg limits such as liquid limit, plastic limit, and plasticity index are important parameters to evaluate soil consistency, strength, and tilth (Larney et al., 1988; Ball et al., 2000; Mueller et al., 2003; Seybold et al., 2008). The Atterberg limits are also used to infer risks of compaction. For example, Proctor maximum bulk density, a measure of maximum soil compactibility, is negatively correlated with liquid and plastic limits (Ball et al., 2000).

The Atterberg limits are positively correlated with SOC (Ball et al., 2000; Mueller et al., 2003). Across 129 Ap (plow) horizons in Canada, liquid and plastic limits (r = 0.68) were positively correlated with SOC concentration (De Jong et al., 1990). Across 12 sites in Ireland, changes in soil organic matter concentration were strongly correlated (r > 0.80) with the Atterberg limits under sugar beet (*Beta vulgaris* L.) with different types of tillage (Larney et al., 1988). Soil organic C in combination other soil properties such as clay content and cation exchange capacity influence liquid limit and plasticity index (Seybold et al., 2008).

While there is limited information on the specific impacts of crop residue removal on SOC concentration vs. consistency relationships, the above studies suggest that decrease in SOC concentration with residue removal can adversely affect soil consistency. Soils with high SOC concentration are more friable, have better tilth, and are less compactable than soils with low SOC concentration. The SOC increases water content at liquid and plastic limits due to the high surface area of organic particles. Soil organic matter particles in association with the mineral fraction increases the water adsorption capacity of the soil. In some soils, correlations between SOC concentration and soil consistency may be weak due to differences in soil parent material, clay content and mineralogy, and type and nature of organic matter (De Jong et al., 1990; Saxton and Rawls, 2006).

Effect of Organic Carbon on Soil Compaction

Changes in SOC concentration due to management may influence risks of soil compaction. The SOC is a sensitive parameter for predicting bulk density changes as the result of soil compaction (Rawls, 1983; Kay et al., 1997; Benites et al., 2007; Ruehlmann and Körschens, 2009; Table 2–1). Kaur et al. (2002) cited a number of pedotransfer functions for predicting bulk density from changes in SOC concentration. Bulk density may decrease linearly or exponentially with increasing SOC concentration (Rawls, 1983; Kay et al., 1997; Ruehlmann and Körschens, 2009). Across 176 sites in Europe, bulk density, determined by the uniaxial compression test, decreased with an increase in organic matter content in soils with <15% of organic matter concentration (Keller and Håkansson, 2010). Soil organic C often interacts with soil particle-size distribution to influence bulk density (Arvidsson, 1998; Kaur et al., 2002; Benites et al., 2007).

Changes in bulk density and SOC concentration can occur rapidly after removal or addition of crop residues. On a sandy loam in Nigeria, application of rice (*Oryza sativa* L.) straw to a no-till soil at 0, 2, 4, 6, and 12 Mg ha^{-1} increased SOC concentration and reduced bulk density in the 0- to 5-cm depth after one and half years of straw application (Lal et al., 1980; Fig. 2–7A). The increase in SOC concentration with residue addition may not only reduce bulk density but also

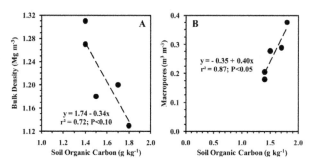

Fig. 2–7. Effect of soil organic C concentration on (A) bulk density and (B) macroporosity after 18 mo of rice straw management (data from Lal et al., 1980).

increase soil macroporosity (Fig. 2–7B). On a loam in Spain, Jordán et al. (2010) reported that applications of 0, 1, 5, 10, and 15 Mg ha^{-1} of wheat straw to a no-till soil for 3 yr reduced bulk density due to an increase in SOC concentration and biological activity. They observed that bulk density decreased from 1.45 to 1.32 Mg m^{-3}, whereas SOC concentration increased from 1 to 117 g kg^{-1} in the 0- to 10-cm depth when 15 Mg ha^{-1} of straw was added. One year after corn stover removal from no-till soils in the eastern United States, a decrease of SOC concentration was significantly and negatively correlated with an increase in bulk density in a sloping silt loam and nearly level clay loam, but the correlation between SOC concentration and bulk density in a nearly level silt loam was not significant in the 0- to 5-cm depth (Table 2–2; Blanco-Canqui et al., 2006a).

Additional crop residue input through intensified cropping systems reduces soil compactibility by increasing SOC concentration. Indeed, soil compactibility as measured by the Proctor test may be one of the most sensitive indicators of tillage-crop-residue management induced changes in SOC concentration (Blanco-Canqui et al., 2009). Proctor maximum bulk density is a measure of maximum soil compactibility and the Proctor critical water content is the soil water content at which the Proctor maximum bulk density occurs. On a loam in Oklahoma, the plow layer of soils under lespedeza (*Lespedeza striata*) had lower Proctor maximum bulk density, a parameter of maximum compactibility of a soil, than that under continuous cotton (*Gossypium hirsutum* L.) due to the greater soil organic matter concentration under lespedeza (Davidson et al., 1967). Across three no-till soils in the central Great Plains, Proctor maximum bulk density was lower in continuous cropping systems than in crop-fallow and crop-crop-fallow systems in two of the three soils by 0.1 to 0.24 Mg m^{-3} in the 0- to 5-cm soil depth (Blanco-Canqui et al., 2010b). The increase in Proctor maximum bulk density was mainly due to a decrease in SOC concentration as it was strongly and negatively correlated (r = –0.84) with SOC in the 0- to 15-cm depth. Increase in SOC concentration with the use of intensive cropping systems can also increase critical water content relative to crop-fallow systems due to differences in annual residue input (Blanco-Canqui et al., 2010b).

Changes in SOC concentration with tillage systems, which dictate rates of residue decomposition, also influence soil compactibility (Table 2–1). We conducted a meta-analysis of the relationships of SOC with Proctor maximum bulk density and Proctor critical water content using data from Wagner et al. (1994), Thomas et al. (1996), Aragón et al. (2000), and Blanco-Canqui et al. (2009) for a wide range of soils managed under no-till, conventional till, and pasture. Results indicated that Proctor maximum bulk density (Fig. 2–8A to 2–8E) and Proctor critical water content (Fig. 2–9A to 2–9E) were strongly correlated with changes in

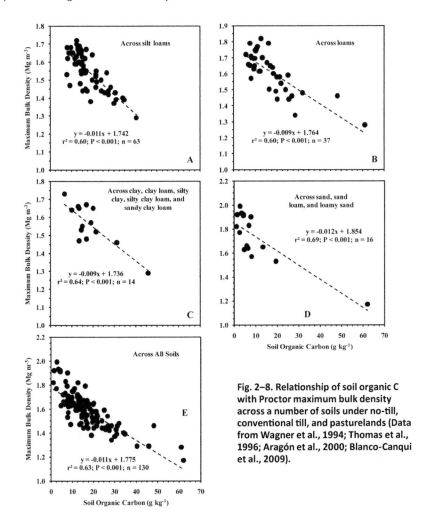

Fig. 2–8. Relationship of soil organic C with Proctor maximum bulk density across a number of soils under no-till, conventional till, and pasturelands (Data from Wagner et al., 1994; Thomas et al., 1996; Aragón et al., 2000; Blanco-Canqui et al., 2009).

SOC concentration. The Proctor maximum bulk density decreased significantly as SOC concentration increased, whereas Proctor critical water content increased as SOC concentration increased. Proctor maximum bulk density and its critical water content were strongly correlated with changes in SOC concentration regardless of differences in soil textural class (Fig. 2–8A to 2–8E and 2–9A to 2–9E) and climatic zones (Fig. 2–10C and 2–10D). These results indicate that a decrease in SOC concentration can increase risks of soil compaction. Soil compactibility is sensitive to management and may be more significantly affected by changes in SOC concentration than other soil physical properties. The decrease in soil water content at which the soil is most compacted due to the decrease in SOC concentration is important to manage soil compaction. The implication is that

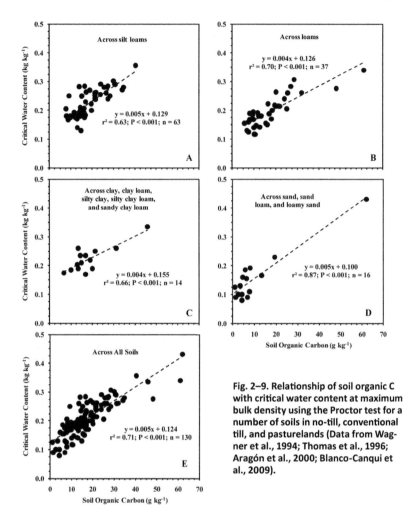

Fig. 2–9. Relationship of soil organic C with critical water content at maximum bulk density using the Proctor test for a number of soils in no-till, conventional till, and pasturelands (Data from Wagner et al., 1994; Thomas et al., 1996; Aragón et al., 2000; Blanco-Canqui et al., 2009).

soils with low SOC concentration can become compacted at lower water content than soils with high SOC concentration under the same traffic load.

The lower soil's susceptibility to compaction under increased SOC concentration is due to the rebounding capacity and low bulk density of organic matter (Soane, 1990). Organic materials are more elastic, porous, and have lower density than inorganic soil particles (Kay, 1997; Rawls, 1983). As result, SOC lowers the bulk density through the "dilution effect" and provides elasticity and rebounding capacity to the whole soil. Soane (1990) used the term "relaxation factor" to explain the spring-like behavior of residue mulch and decomposed organic materials. Soils with high SOC concentration are more resilient and have greater "relaxation factor" than those with low SOC concentration. Organic matter

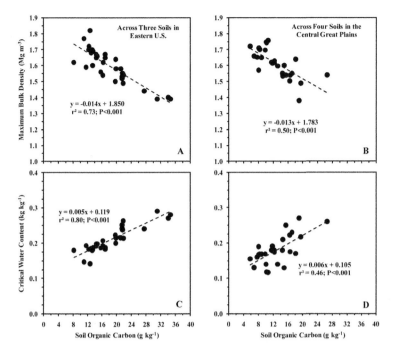

Fig. 2–10. Effect of soil organic C on (A and B) Proctor maximum bulk density and (C and D) critical water content at maximum compaction in the 0- to 5-cm soil depth in two different climatic regions in the United States (data from Thomas et al., 1996; Blanco-Canqui et al., 2009).

also changes the strength of bonds and electrical charges at the intra-aggregate contact points between organic and inorganic particles, which can change the behavior of the soil matrix (Soane, 1990; Ball et al., 2000).

The increase in maximum compactibility with decreased SOC concentration can have important implications for managing crop residues and soil compaction. It suggests that residues should be returned to soil to maintain or increase SOC and to reduce, at least in part, some of the risks of excessive compaction. The focus to alleviate soil compaction has been on reducing axle loads, controlling timing and frequency of traffic, and implementing remediation measures such as subsoiling, vertical tillage, and others. The ability of SOC to influence the soils' susceptibility to compaction has been somewhat ignored when managing excessive soil compaction. While crop residue mulch alone may not be highly effective in reducing soil bulk density from an increase in applied stress (Gupta et al., 1987), SOC accumulation with continued residue addition may improve soil resilience and rebounding capacity in the long term. The role of SOC in alleviating excessive soil compaction can be particularly relevant at low than at high axle loads of field equipment. Overall, because SOC management is critical to reduce soil compactibility, reduction in SOC concentration by practices such as residue removal can reduce the soils' ability to buffer risks of excessive compaction.

Effect of Organic Carbon on Soil Hydraulic Properties

The adverse impacts of SOC loss due to crop residue removal on soil compaction parameters and structural stability can concomitantly result in degraded soil hydraulic properties (Table 2–1). Soil organic C loss can reduce soil macro- and microporosity, water retention, and saturated and unsaturated hydraulic conductivity (Rawls et al., 2004). The increased bulk density and reduced aggregate stability with reduced SOC concentration can alter pore-size distribution and reduce soil macroporosity. Residue-derived SOC interacts with soil particles and alters both soil matrix and structural architecture, thereby modifying pore characteristics and increasing specific surface area of the soil.

Figures 2–11A to 2–11C show that an increase in SOC concentration with residue return increases soil porosity. Improved soil porosity increases rates of water, air, and heat flow through the soil. Input of SOC-enriched residues also stimulates activity of soil organisms (e.g., earthworms), which create water-conducting biopores. Weak aggregates with low SOC concentration near the soil surface are prone to slaking, which seals the soil surface, and causes closure of open-ended macropores, thereby reducing water infiltration and inducing surface runoff (Rhoton et al., 2002). Crop-residue-derived SOC with slight hydrophobic properties can reduce aggregate slaking and maintain the integrity of water conducting macropores. Water-repellent organic compounds coat macropore walls and influence water flow (Wang et al., 2009). Using shrinkage curve analysis, Boivin et al. (2009) reported that an increase in SOC concentration linearly increased soil

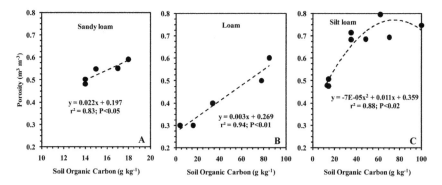

Fig. 2–11. Effect of soil organic C on total porosity for three soils under different scenarios of residue management. Data from Lal et al. (1980) for (A) the sandy loam in the 0- to 5-cm soil depth, Jordán et al. (2010) for (B) the loam in the 0- to 10-cm soil depth, and Blanco-Canqui et al. (2007) for (C) the silt loam in the 0- to 7.5-cm soil depth were used.

pore volume, water retention, and soil structural parameters, and changes in SOC concentration had much stronger effects than changes in clay concentration.

Differences in SOC concentration may also affect porosity by altering soil particle density (Table 2–1). The few studies have available on this topic have shown that particle density decreases with an increase in SOC concentration. Across various cultivated soils in the UK, Ball et al. (2000) reported that particle density was negatively and significantly correlated ($r = -0.38$; $P < 0.001$) with SOC concentration. Similarly, across no-till, chisel plow, and moldboard plow systems in Ohio, Blanco-Canqui et al. (2006b) found that particle density was as sensitive to changes in SOC concentration as bulk density. They reported that a decrease in SOC concentration due to differences in residue management between no-till and plowed systems explained 38% ($P < 0.001$) of the variability in particle density in the 0- to 10-cm soil depth. Similar to the effects on bulk density, the decrease in particle density with increase in SOC concentration is attributed to the dilution effect of soil organic particles. Changes in particle density can affect soil hydraulic properties by altering soil porosity.

Many studies have shown that residue management-induced changes in SOC concentration alter water retention. On two silt loams in Iowa, increased SOC concentration by doubling the amount of corn stover for 10 yr in no-till increased plant available water at −0.5, −1.4, and −9.8 kPa (Karlen et al., 1994). On a silt loam in Ohio, wheat straw addition to no-till plots for 7 yr increased both water retention at >30 kPa suctions and SOC concentration in the 0- to 10-cm depth (Duiker and Lal, 1999). Correlations in Table 2–2 for three contrasting no-till soils show that water retention and plant available water decreased linearly with a loss in SOC concentration due to corn stover removal (Blanco-Canqui et al., 2006a; Blanco-Canqui et al., 2007). Decrease in SOC concentration reduces the soil's ability to absorb and retain water because it reduces the specific surface area of the soil. Organic particles have a greater specific surface area and water adsorption capacity than soil inorganic particles (Rawls et al., 2003). Hudson (1994) found that soils containing 4% organic matter retained plant available water twice more than soils containing 1% organic matter. Olness and Archer (2005) found that change in plant available water ranged between 2.5 and 5% for each 1% change in SOC concentration for soils with <2.5% SOC and 40% clay concentrations. Recently, Kvaerno and Haugen (2011), while assessing the performance of a number of pedotransfer functions in predicting soil water characteristics based on particle-size distribution, organic matter content, and bulk density across 540 soil horizons on cultivated lands in Norway, found that pedotransfer functions which included organic matter content as one of the input parameters were the best predictors of soil water retention under low suctions.

Interactions among soil water, water repellency, aggregate hierarchy, SOC, and pore-size distribution are the mechanisms that determine water transmission and retention characteristics. For example, soil macroaggregates influence soil water retention differently from microaggregates because of differences in intra-aggregate pore size and the amount of soil organic material found inside the pores. At the microstructural level, microaggregates influence soil water retention through reduced soil pore connectivity and increased water retention in micropores. Zhuang et al. (2008) observed that soil organic matter inside <5 µm diam. pores increased soil water retention in microaggregates. They also observed that micropores filled with soil organic matter caused hysteresis of soil water retention relative to micropores with low concentration of organic matter. The implication is that soil micropores occluding organic matter (e.g., recalcitrant SOC) can retain more water for longer periods of time than micropores filled with little or no organic matter. Protection of SOC inside of aggregates not only is essential for long-term SOC sequestration but also for improvement in soil water retention capacity (Zhuang et al., 2007).

It is important to note that relationship between SOC concentration and soil hydraulic properties can be complex and not always consistent (Olness and Archer, 2005; Benjamin et al., 2008; Wang et al., 2009). For example, Rawls et al. (2004) observed that at low SOC concentrations, water retention was negatively correlated with SOC concentration in clayey soils. In some sandy soils with low SOC concentration, SOC-induced increase in soil water repellency may reduce saturated hydraulic conductivity, offsetting any positive effects of improved soil aggregation induced by SOC on saturated hydraulic conductivity (Wang et al., 2009). Impacts of SOC concentration on soil water retention capacity may be greater than on saturated hydraulic conductivity (Karlen et al., 1994; Nemes et al., 2005; Benjamin et al., 2008). Karlen et al. (1994) reported that decreased SOC concentration with residue removal reduced soil water retention capacity at lower suctions, but it did not affect saturated hydraulic conductivity. Rawls et al. (2004) reported, in contrast, that an increase in SOC concentration increased saturated hydraulic conductivity when the clay proportion was <0.3, but the impact of SOC was not clear for clay proportion was >0.3.

Conclusions

Changes in SOC concentration due to crop residue management are related to changes in soil physical behavior. The magnitude at which changes in SOC concentration affect soil physical behavior are nevertheless affected by the amount and constituents of SOC, tillage and cropping system, and soil textural class, among others. The SOC interacts with soil intrinsic properties such as particle-size distribution to influence soil compaction, structural, and hydraulic properties. The SOC buffers risks of excessive soil compaction, increases soil aggregate stability and strength, promotes macroporosity, induces slight water repellency, and improves water retention.

The mechanisms by which SOC influences soil physical properties are numerous and complex. Organic particles stabilize soil aggregates by binding individual particles into stable units and strengthening the inter-particle cohesion within and among aggregates. Organic films can also induce some hydrophobic properties to soil, reducing aggregate slaking. Because crop residues have elastic properties, residue-derived organic materials provide elasticity, spring-like behavior, and rebounding capacity to the whole soil. Organic particles also have lower density than mineral particles, which dilutes the soil bulk density, reducing risks of excessive compression and compaction of the soil. Presence of a network of fine roots, fungal hyphae, and other biological components can enmesh mineral particles and increase friction forces among soil particles. Organic particles can also impart slight electrical charge to the soil similar to clay particles to react and develop complex chemical bonds among soil particles to further improve soil physical properties. These myriad benefits of SOC-enriched materials can be readily altered by management practices such as crop residue removal.

Crop residue removal adversely impacts soil physical properties by depleting SOC, but C input through high-biomass producing crop rotations (e.g., continuous cropping systems) may maintain and improve soil physical characteristics. Residue management strategies (e.g., no-till) that increase SOC concentration improve soil structural, compaction, and hydraulic properties. Particularly, an increase in SOC concentration is strongly correlated with maximum soil compactibility and critical water content, indicating that cultivated soils with increased SOC concentration are less susceptible to compaction and can be trafficked at greater soil water content without the risks of soil compaction compared with soils low in SOC concentration.

The numerous beneficial effects of SOC on different soil physical parameters support the need for maintaining optimum levels of SOC through annual crop residue return and use of no-till farming to maintain or improve soil functions. Because excessive removal of crop residues for off-farm uses readily reduces SOC concentration, it can adversely affect soil physical behavior. Residue mulch improves soil physical properties not only by increasing SOC concentration but also by protecting the soil surface from the erosive forces of raindrops, and reducing abrupt fluctuations of soil temperature, freezing and thawing, and wetting and drying cycles. Overall, increasing SOC concentration through proper crop residue management may not only reduce net emissions of C to the atmosphere and contribute to nutrient cycling but can also improve soil physical properties. More research documenting the independent effects of crop residue management on the relationships between residue-derived SOC and soil physical properties is needed for different soil types, tillage and cropping systems, and climatic conditions. This information is important to better understand the implications of crop residue removal or addition on soil physical properties as result of changes in SOC concentration. Research is also needed to identify the specific pools or constituents of SOC that are more related to changes in soil physical properties relative to total SOC.

References

Anderson-Teixeira, K.J., S.C. Davis, M.D. Masters, and E.H. Delucia. 2009. Changes in soil organic carbon under biofuel crops. GCB Bioenergy 1:75–96. doi:10.1111/j.1757-1707.2008.01001.x.

Aragón, A., M.G. Garcia, R.R. Filgueira, and Y.A. Pachepsky. 2000. Maximum compactibility of Argentine soils from the Proctor test: The relationship with organic carbon and water content. Soil Tillage Res. 56:197–204. doi:10.1016/S0167-1987(00)00144-6.

Arvidsson, J. 1998. Influence of soil texture and organic matter content on bulk density, air content, compression index and crop yield in field and laboratory compression experiments. Soil Tillage Res. 49:159–170. doi:10.1016/S0167-1987(98)00164-0.

Ball, B.C., D.J. Campbell, and E.A. Hunter. 2000. Soil compactibility in relation to physical and organic properties at 156 sites in UK. Soil Tillage Res. 57:83–91. doi:10.1016/S0167-1987(00)00145-8.

Ball, B.C., R.W. Lang, M.F. O'Sullivan, and M.F. Franklin. 1989. Cultivation and nitrogen requirements for continuous winter barley on a Gleysol and a Cambisol. Soil Tillage Res. 13:333–352. doi:10.1016/0167-1987(89)90042-1.

Bell, M.J., P.W. Moody, R.D. Connolly, and B.J. Bridge. 1998. The role of active fractions of soil organic matter in physical and chemical fertility of Ferrosols. Aust. J. Soil Res. 36:809–819. doi:10.1071/S98020.

Benites, V.M., P.O.O.A. Machado, E.C.C. Fidalgo, M.R. Coelho, and B.E. Madari. 2007. Pedotransfer functions for estimating soil bulk density from existing soil survey reports in Brazil. Geoderma 139:90–97. doi:10.1016/j.geoderma.2007.01.005.

Benjamin, J.G., A.D. Halvorson, D.C. Nielsen, and M.M. Mikha. 2010. Crop management effects on crop residue production and changes in soil organic carbon in the central Great Plains. Agron. J. 102:990–997. doi:10.2134/agronj2009.0483.

Benjamin, J.G., M.M. Mikha, and M.F. Vigil. 2008. Organic carbon effects on soil physical and hydraulic properties in a semiarid climate. Soil Sci. Soc. Am. J. 72:1357–1362. doi:10.2136/sssaj2007.0389.

Bhogal, A., F.A. Nicholson, and B.J. Chambers. 2009. Organic carbon additions: Effects on soil bio-physical and physicochemical properties. Eur. J. Soil Sci. 60:276–286. doi:10.1111/j.1365-2389.2008.01105.x.

Blanco-Canqui, H. 2011. Does no-till farming induce water repellency to soils? Soil Use Manage. 27:2–9. doi:10.1111/j.1475-2743.2010.00318.x.

Blanco-Canqui, H., J.G. Benjamin, A. J. Schlegel, L.R. Stone, and P.W. Stahlman. 2010b. Continuous cropping systems reduce near-surface compaction in no-till soils. Agron. J. 102:1217–1225. doi:10.2134/agronj2010.0113.

Blanco-Canqui, H., and R. Lal. 2007. Soil structure and organic carbon relationships following 10 years of wheat residue management. Soil Tillage Res. 95:240–254. doi:10.1016/j.still.2007.01.004.

Blanco-Canqui, H., and R. Lal. 2008a. Crop residue management and soil carbon dynamics. In: R. Lal, editor, Soil carbon sequestration and the greenhouse effect. SSSA Spec. Publ. 57, SSSA, Madison, WI.

Blanco-Canqui, H., and R. Lal. 2008b. Corn stover removal impacts on micro-scale soil physical properties. Geoderma 145:335–346. doi:10.1016/j.geoderma.2008.03.016.

Blanco-Canqui, H., R. Lal, R.C. Izaurralde, W.M. Post, and M.J. Shipitalo. 2006b. Organic carbon influences on particle density and rheological properties for a silt loam soil. Soil Sci. Soc. Am. J. 70:1407–1414. doi:10.2136/sssaj2005.0355.

Blanco-Canqui, H., R. Lal, L.B. Owens, W.M. Post, and M.J. Shipitalo. 2007. Rapid changes in soil carbon and structural properties due to stover removal from no-till corn plots. Soil Tillage Res. 92:144–155. doi:10.1016/j.still.2006.02.002.

Blanco-Canqui, H., R. Lal, W.M. Post, R.C. Izaurralde, and L.B. Owens. 2006a. Soil structural parameters and organic carbon in no-till corn with variable stover retention rates. Soil Sci. 171:468–482. doi:10.1097/01.ss.0000209364.85816.1b.

Blanco-Canqui, H., L.R. Stone, A.J. Schlegel, D.J. Lyon, M.F. Vigil, M. Mikha, and P.W. Stahlman. 2009. No-till induced increase in organic carbon reduces maximum bulk density of soils. Soil Sci. Soc. Am. J. 73:1871–1879. doi:10.2136/sssaj2008.0353.

Blanco-Canqui, H., L.R. Stone, and P.W. Stahlman. 2010a. Soil response to long-term cropping systems on an Argiustoll in the central Great Plains. Soil Sci. Soc. Am. J. 74:602–611. doi:10.2136/sssaj2009.0214.

Boivin, P., B. Schäffer, and W. Sturny. 2009. Quantifying the relationship between soil organic carbon and soil physical properties using shrinkage modelling. Eur. J. Soil Sci. 60:265–275. doi:10.1111/j.1365-2389.2008.01107.x.

Bordovsky, D.G., M. Choudhary, and C.J. Gerard. 1999. Effect of tillage, cropping, and residue management on soil properties in the Texas Rolling Plains. Soil Sci. 164:331–340. doi:10.1097/00010694-199905000-00005.

Bossuyt, H., J. Six, and P.F. Hendrix. 2005. Protection of soil carbon by microaggregates within earthworm casts. Soil Biol. Biochem. 37:251–258. doi:10.1016/j.soilbio.2004.07.035.

Bottinelli, N., V. Hallaire, S. Menasseri-Aubry, C. Le Guillou, and D. Cluzeau. 2010. Abundance and stability of belowground earthworm casts influenced by tillage intensity and depth. Soil Tillage Res. 106:263–267. doi:10.1016/j.still.2009.11.005.

Chaney, K., and R.S. Swift. 1984. The influence of organic matter on aggregate stability in some British soils. J. Soil Sci. 35:223–230. doi:10.1111/j.1365-2389.1984.tb00278.x.

Chenu, C., Y. Le Bissonnais, and D. Arrouays. 2000. Organic matter influence on clay wettability and soil aggregate stability. Soil Sci. Soc. Am. J. 64:1479–1486. doi:10.2136/sssaj2000.6441479x.

Davidson, J.M., F. Gray, and D.I. Pinson. 1967. Changes in organic matter and bulk density with depth under two cropping systems. Agron. J. 59:375–378. doi:10.2134/agronj1967.00021962005900040025x.

De Jong, E., D.F. Acton, and H.B. Stonehouse. 1990. Estimating the Atterberg limits of southern Saskatchewan soils from texture and carbon contents. Can. J. Soil Sci. 70:543–554. doi:10.4141/cjss90-057.

De Jonge, L.W., P. Moldrup, and O.H. Jacobsen. 2007. Soil-water content dependency of water repellency in soils: Effect of crop type, soil management, and physical-chemical parameters. Soil Sci. 172:577–588. doi:10.1097/SS.0b013e318065c090.

Dexter, A.R., G. Richard, D. Arrouays, E.A. Czyz, C. Jolivet, and O. Duval. 2008. Complexed organic matter controls soil physical properties. Geoderma 144:620–627. doi:10.1016/j.geoderma.2008.01.022.

Diaz-Zorita, M., and G.A. Grosso. 2000. Effect of soil texture, organic carbon and water retention on the compactibility of soils from the Argentinean pampas. Soil Tillage Res. 54:121–126. doi:10.1016/S0167-1987(00)00089-1.

Doerr, S.H., R.A. Shakesby, and R.P.D. Walsh. 2000. Soil water repellency: Its causes, characteristics and hydro-geomorphological significance. Earth Sci. Rev. 51:33–65. doi:10.1016/S0012-8252(00)00011-8.

Duiker, S.W., and R. Lal. 1999. Crop residue and tillage effects on C sequestration in a Luvisol in central Ohio. Soil Tillage Res. 52:73–81. doi:10.1016/S0167-1987(99)00059-8.

Eynard, A., T.E. Schumacher, M.J. Lindstrom, D.D. Malo, and R.A. Kohl. 2006. Effects of aggregate structure and organic C on wettability of Ustolls. Soil Tillage Res. 88:205–216. doi:10.1016/j.still.2005.06.002.

Goebel, M.-O., J. Bachmann, S.K. Woche, and W.R. Fischer. 2005. Soil wettability, aggregated stability, and the decomposition of soil organic matter. Geoderma 128:80–93. doi:10.1016/j.
geoderma.2004.12.016.

Gupta, S.C., E.C. Schneider, W.E. Larson, and A. Hadas. 1987. Influence of corn residue on compression and compaction behavior of soils. Soil Sci. Soc. Am. J. 51:207–212. doi:10.2136/sssaj198
7.03615995005100010043x.

Hallett, P.D., T. Baumgartl, and I.M. Young. 2001. Subcritical water repellency of aggregates
from a range of soil management practices. Soil Sci. Soc. Am. J. 65:184–190. doi:10.2136/
sssaj2001.651184x.

Haynes, R.J. 2005. Labile organic matter fractions as central components of the quality of agricultural soils: An overview. Adv. Agron. 85:221–268. doi:10.1016/S0065-2113(04)85005-3.

Hudson, B.D. 1994. Soil organic matter and available water capacity. J. Soil Water Conserv.
49:189–194.

Imhoff, S., A. Pires da Silva, and A. Dexter. 2002. Factors contributing to the tensile strength and friability of Oxisols. Soil Sci. Soc. Am. J. 66:1656–1661. doi:10.2136/sssaj2002.1656.

Johnson, J.M.F., D. Reicosky, B. Sharratt, M. Lindstrom, W. Voorhees, and L. Carpenter-Boggs. 2004.
Characterization of soil amended with the by-product of corn stover fermentation. Soil Sci.
Soc. Am. J. 68:139–147.

Jordán, A., L.M. Zavala, and J. Gil. 2010. Effects of mulching on soil physical properties and
runoff under semi-arid conditions in southern Spain. Catena 81:77–85. doi:10.1016/j.
catena.2010.01.007.

Karlen, D.L., N.C. Wollenhaupt, D.C. Erbach, E.C. Berry, J.B. Swan, N.S. Eash, and J.L. Jordahl. 1994.
Crop residue effects on soil quality following 10 years of no-till corn. Soil Tillage Res. 31:149–
167. doi:10.1016/0167-1987(94)90077-9.

Kay, B.D. 1997. Soil structure and organic C: A review. In: R. Lal et al., editors, Soil processes and the
carbon cycle. CRC Press, Boca Raton, FL. p. 169–197.

Kay, B.D., A.P. da Silva, and J.A. Baldock. 1997. Sensitivity of soil structure to changes in organic
carbon content: Predictions using pedotransfer functions. Can. J. Soil Sci. 77:655–667.
doi:10.4141/S96-094.

Kaur, R., S. Kumar, and H.P. Gurung. 2002. A pedo-transfer function (PTF) for estimating soil bulk
density from basic soil data and its comparison with existing PTFs. Aust. J. Soil Res. 40:847–
857. doi:10.1071/SR01023.

Keller, T., and I. Håkansson. 2010. Estimation of reference bulk density from soil particle size
distribution and soil organic matter content. Geoderma 154:398–406. doi:10.1016/j.
geoderma.2009.11.013.

Kvaerno, S.H., and L.E. Haugen. 2011. Performance of pedotransfer functions in predicting soil
water characteristics of soils in Norway. Acta Agric. Scand. Sec. B 61:264–280.

Lal, R. 1997. Residue management, conservation tillage and soil restoration for mitigat-
ing greenhouse effect by CO_2-enrichment. Soil Tillage Res. 43:81–107. doi:10.1016/
S0167-1987(97)00036-6.

Lal, R., D. DeYleeschauwer, and R.M. Nganje. 1980. Changes in properties of newly cleared Alfisol
as affected by mulching. Soil Sci. Soc. Am. J. 44:827–833. doi:10.2136/sssaj1980.036159950
04400040034x.

Lamparter, A., J. Bachmann, M.O. Goebel, and S.K. Woche. 2009. Carbon mineralization in soil:
Impact of wetting-drying, aggregation and water repellency. Geoderma 150:324–333.
doi:10.1016/j.geoderma.2009.02.014.

Larney, F.J., R.A. Fortune, and J.F. Collins. 1988. Intrinsic soil physical parameters influencing
intensity of cultivation procedures for sugar-beet seedbed preparation. Soil Tillage Res.
12:253–267. doi:10.1016/0167-1987(88)90015-3.

Liebig, M.A., G.E. Varvel, J.W. Doran, and B.J. Wienhold. 2002. Crop sequence and nitrogen fertilization effects on soil properties in the Western Corn Belt. Soil Sci. Soc. Am. J. 66:596–601.
doi:10.2136/sssaj2002.0596.

Liu, A., B.L. Ma, and A.A. Bomke. 2005. Effects of cover crops on soil aggregate stability, total
organic carbon, and polysaccharides. Soil Sci. Soc. Am. J. 69:2041–2048. doi:10.2136/
sssaj2005.0032.

MacDonald, L.H., and E.L. Huffman. 2004. Post-fire soil water repellency: Persistence and soil moisture thresholds. Soil Sci. Soc. Am. J. 68:1729–1734. doi:10.2136/sssaj2004.1729.

Malamoud, K., A.B. McBratney, B. Minasny, and D.J. Field. 2009. Modelling how carbon affects soil structure. Geoderma 149:19–26. doi:10.1016/j.geoderma.2008.10.018.

Mueller, L., U. Schindler, N.R. Fausey, and R. Lal. 2003. Comparison of methods for estimating maximum soil water content for optimum workability. Soil Tillage Res. 72:9–20. doi:10.1016/S0167-1987(03)00046-1.

Nemes, A., W.J. Rawls, and Y.A. Pachepsky. 2005. Influence of organic matter on the estimation of saturated hydraulic conductivity. Soil Sci. Soc. Am. J. 69:1330–1337. doi:10.2136/sssaj2004.0055.

Olness, A., and D. Archer. 2005. Effect of organic carbon on available water in soil. Soil Sci. 170:90–101. doi:10.1097/00010694-200502000-00002.

Perlack, R.D., L.L. Wright, A.F. Turhollow, R.L. Graham, B.J. Stokes, and D.C. Erbach. 2005. Biomass as feedstock for a bioenergy and bioproducts industry: The technical feasibility of a billion-ton annual supply. Rep. ORLN/TM-2005/66. Oak Ridge Natl. Lab., Oak Ridge, TN.

Pikul, J.L., G. Chilom, J. Rice, A. Eynard, T.E. Schumacher, K. Nichols, J.M.F. Johnson, S. Wright, T. Caesar, and M. Ellsbury. 2009. Organic matter and water stability of field aggregates affected by tillage in South Dakota. Soil Sci. Soc. Am. J. 73:197–206. doi:10.2136/sssaj2007.0184.

Puget, P., R. Lal, C. Izaurralde, M. Post, and L. Owens. 2005. Stock and distribution of total and corn-derived soil organic carbon in aggregate and primary particle fractions for different land use and soil management practices. Soil Sci. 170:256–279. doi:10.1097/00010694-200504000-00004.

Quiroga, A.R., D.E. Buschiazzo, and N. Peinemann. 1999. Soil compaction is related to management practices in the semi-arid Argentine pampas. Soil Tillage Res. 52:21–28. doi:10.1016/S0167-1987(99)00049-5.

Rawls, W.J. 1983. Estimating soil bulk density from particle size analysis and organic matter content. Soil Sci. 135:123–125. doi:10.1097/00010694-198302000-00007.

Rawls, W.J., Y.A. Pachepsky, J.C. Ritchie, T.M. Sobecki, and H. Bloodworth. 2003. Effect of soil organic carbon on soil water retention. Geoderma 116:61–76. doi:10.1016/S0016-7061(03)00094-6.

Rawls, W.J., A. Nemes, and Y.A. Pachepsky. 2004. Effect of soil organic carbon on soil hydraulic properties. In: Development of pedotransfer functions in soil hydrology. Development in soil sci. 30. Elsevier, New York. p. 95–111.

Reddy, N., and Y. Yang. 2005. Biofibers from agricultural byproducts for industrial applications. Trends Biotechnol. 23:22–27. doi:10.1016/j.tibtech.2004.11.002.

Rhoton, F.E., M.J. Shipitalo, and D.L. Lindbo. 2002. Runoff and soil loss from midwestern and southeastern US silt loam soils as affected by tillage practice and soil organic matter content. Soil Tillage Res. 66:1–11. doi:10.1016/S0167-1987(02)00005-3.

Rodriguez-Alleres, M., E. Benito, and E. de Blas. 2007. Extent and persistence of water repellency in north-western Spanish soils. Hydrol. Processes 21:2291–2299. doi:10.1002/hyp.6761.

Ruehlmann, J., and M. Körschens. 2009. Calculating the effect of soil organic matter concentration on soil bulk density. Soil Sci. Soc. Am. J. 73:876–885. doi:10.2136/sssaj2007.0149.

Saxton, K.E., and W.J. Rawls. 2006. Soil water characteristic estimates by texture and organic matter for hydrologic solutions. Soil Sci. Soc. Am. J. 70:1569–1578. doi:10.2136/sssaj2005.0117.

Seybold, C.A., M.A. Elrashidi, and R.J. Engel. 2008. Linear regression models to estimate soil liquid limit and plasticity index. Soil Sci. 173:25–34. doi:10.1097/ss.0b013e318159a5e1.

Singh, B., D.S. Chanasyk, W.B. Mcgill, and M.P.K. Nyborg. 1994. Residue and tillage management effects on soil properties of a Typic Cryoboroll under continuous barley. Soil Tillage Res. 32:117–133. doi:10.1016/0167-1987(94)90015-9.

Soane, B.D. 1990. The role of organic matter in soil compactability: A review of some practical aspects. Soil Tillage Res. 16:179–201. doi:10.1016/0167-1987(90)90029-D.

Spaccini, R., A. Piccolo, J.S.C. Mbagwu, A.Z. Teshale, and C.A. Igwe. 2002. Influence of the addition of organic residues on carbohydrate content and structural stability of some highland soils in Ethiopia. Soil Use Manage. 18:404–411. doi:10.1079/SUM2002152.

Sparrow, S.D., C.E. Lewis, and C.W. Knight. 2006. Soil quality response to tillage and crop residue removal under subarctic conditions. Soil Tillage Res. 91:15–21. doi:10.1016/j.still.2005.08.008.

Tanaka, D.L., J.F. Karn, M.A. Liebig, S.L. Kronberg, and J.D. Hanson. 2005. An integrated approach to crop/livestock systems: Forage and grain production for swath grazing. Renew. Agric. Food Syst. 20:223–231. doi:10.1079/RAF2005107.

Thomas, G.W., G.R. Hazler, and R.L. Blevins. 1996. The effects of organic matter and tillage on maximum compactibility of soils using the Proctor test. Soil Sci. 161:502–508. doi:10.1097/00010694-199608000-00005.

Tisdall, J.M., and J.M. Oades. 1982. Organic matter and water stable aggregates in soils. J. Soil Sci. 33:141–163. doi:10.1111/j.1365-2389.1982.tb01755.x.

Villamil, M.B., G.A. Bollero, R.G. Darmody, F.W. Simmons, and D.G. Bullock. 2006. No-till corn/soybean systems including winter cover crops: Effects on soil properties. Soil Sci. Soc. Am. J. 70:1936–1944. doi:10.2136/sssaj2005.0350.

Wagner, L.E., N.M. Ambe, and D. Ding. 1994. Estimating a Proctor density curve from intrinsic soil properties. Trans. ASAE 37:1121–1125.

Wang, T., D.A. Wedin, and V.A. Zlotnik. 2009. Field evidence of a negative correlation between saturated hydraulic conductivity and soil carbon in a sandy soil. Water Resour. Res. 45:W00A11. doi:10.1029/2007WR006768.

Weil, R.R., and F. Magdoff. 2004. Significance of soil organic matter to soil quality and health. In: F.R. Magdoff and R.R. Weil, editors, Soil organic matter in sustainable agriculture. CRC Press, New York. p. 1–43.

Wilhelm, W.W., J.M.F. Johnson, J.L. Hatfield, W.B. Voorhees, and D.R. Linden. 2004. Crop and soil productivity response to corn residue removal: A literature review. Agron. J. 96:1–17. doi:10.2134/agronj2004.0001.

Zhuang, J., J.F. McCarthy, E. Perfect, L.M. Mayer, and J.D. Jastrow. 2008. Soil water hysteresis in water-stable microaggregates as affected by organic matter. Soil Sci. Soc. Am. J. 72:212–220. doi:10.2136/sssaj2007.0001S6.

Chapter 3

High Energy Moisture Characteristics: Linking Between Some Soil Physical Processes and Structure Stability

A.I. Mamedov and G.J. Levy

Abstract

Soil structure is the combination or arrangement of primary soil particles into secondary units that are characterized on the basis of size, shape, and grade. Stability of soil structure describes the ability of the soil to retain its arrangement of solid and pore space (i.e., aggregates and pores) when exposed to external forces (e.g., tillage, cropping, compaction and irrigation). Studying the effects of agricultural management practices on soil structure is important for the development of effective soil conservation practices to avoid risks of soil deterioration. The current review presents and discusses results of water retention curves at near saturation (matric potential, ψ, 0 to −5 J kg^{-1}) obtained by the High Energy Moisture Characteristic (HEMC) method for >200 soil samples from humid and arid zones. Quantifying differences in the water retention curves using the modified van Genuchten model (Soil-HEMC) yielded soil structure indices which were then used to elucidate in quantitative terms possible interactions among soil physical processes and management practices. Analyses of the water retention curves indicated that kaolinitic humid zone soils are (i) more stable than smectitc semiarid zone soils, and (ii) less sensitive to management, soil texture, organic matter content and wetting rate. Parameters of the van Genuchten model were successfully employed for evaluation of the contribution of large (>0.25–0.50 mm) and small (<0.25–0.50 mm) aggregates/particles to the pore size distribution and soil structure condition. The presented detailed analyses of the contribution of soil inherent properties and extrinsic conditions (e.g., spatiotemporal variation) to inter- and intra-aggregate porosity, the soil structure indices and model parameters at near saturation, sheds additional light on the processes that are involved in stabilizing soil structure, and may assist in developing more efficient soil and environment management practices.

Abbreviations: ESP, exchangeable sodium percentage; HEMC, high energy moisture characteristic; MS, modal suction; PSD, pore size distribution; SAR, sodium adsorption ratio; SI, structural index; SR, stability ratio; VDP, volume of drainable pores.

A.I. Mamedov, USDA-ARS, Engineering and Wind Erosion Research Unit, 1515 College Ave., Kansas State Univ., Manhattan, KS 66502 (am03@ksu.edu); G.J. Levy, Institute of Soil, Water and Environmental Sciences, Agricultural Research Organization, POB 6, Bet Dagan 50250, Israel (vwguy@volcani.agri.gov.il).

doi:10.2134/advagricsystmodel3.c3

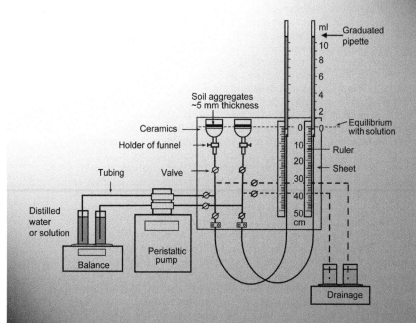

Soil Structure and Aggregate Breakdown Mechanisms

Soil structure is the combination or arrangement of primary soil particles into secondary units. The secondary units are characterized on the basis of size, shape, and grade (SSSA internet glossary, https://www.soils.org/publications/soils-glossary). Structural form of soils refers to total porosity, size distribution and continuity of the pore system and therefore, describes the arrangement and size of inter-and intra-aggregate pores. Stability of soil structure describes the ability of the soil to retain its arrangement of solid and pore space (i.e., aggregates and pores) when exposed to external forces (e.g., tillage, cropping, compaction and irrigation). The soil structural condition and stability are critical for understanding (i) the influence of structural porosity on the nutrient regime in the root zone and thus on crop yield, (ii) soil hydraulic properties and transport processes in macropores, both in the root and the vadose zone (Logsdon and Jaynes, 1993; Shouse and Mohanty, 1998; Vogel and Roth, 1998; Angulo-Jaramillo et al., 2000; Logsdon, 2002; Or and Ghezzehei, 2002; Green et al., 2003; Malone et al., 2003; Kutílek, 2004; Assouline, 2006; Ahuja et al., 2006; Jarvis, 2007).

Studying the effects of agricultural management practices on soil structure is important for the development of effective soil conservation practices to avoid risks of soil deterioration. However, although soil structure a vital soil property which affects several processes important to soils' behavior (e.g., water storage, drainage ability, and runoff generation), its productive capacity and environmental quality, soil structure is often treated as a qualitative and subjective concept, probably because direct measurement of soil structure is complicated and there is no single approach that can meet all the needed requirements for soil structure assessment (Lal, 1991). Tillage and subsequent soil reconsolidation can change soil bulk density and porosity and subsequently affect soils' physical and chemical properties, and nutrients availability. Furthermore after tillage and during the growing season, plant growth and microorganisms' activities interact with environmental variables such as dry–wet and freeze–thaw cycles to modify and affect soil structure (e.g., review papers by Collis-George, 1991; Sumner and Stewart, 1992; Horn et al.,1994; Connolly, 1998; Angers and Caron, 1998; Haynes and Naidu, 1998; Kay and Angers, 2002; Hamza and Anderson, 2005; Ahuja et al., 2006; Peigne et al., 2007; Strudley et al., 2008).

The observed differences in soil physical properties among management practices, such as tillage and cropping systems, are usually transient because large pores created by tillage often collapse after rainfall impact and drying cycles (van Es, 1993; Ahuja et al., 1998; Karlen et al., 2006; Strudley et al., 2008; Benjamin et al., 2008). The ability to study soil structure dynamics and affecting mechanisms thereon are complicated by the following factors (e.g., Strudley et al., 2008): (i) the effects of long-term management (e.g., no-till, crop rotation, crop residues) on soil structure (macropores and macropore continuity) and hydraulic properties are frequently unclear and less evident than those of the original or natural soil properties, (ii) the magnitude and pattern of temporal variability are affected considerably by the spatial location and growing season, (iii) the spatial and temporal variability often overshadow the effects of particular management practices, (iv) to the difficulty involved in relating results from laboratory measurements of soil structure stability and pore-size distribution (PSD) to actual field behavior in terms of infiltration, tillage requirements and productivity, and (v) soil types and climate affect the resultant soil structure and soil hydraulic properties, thus transfer of effects of a given treatment from one location to another could prove problematic. Therefore, the limited understanding of soil structure led to the use of a flawed approach in many water and solute flow, runoff generation and chemical transfer studies and models, where by the soil is considered as a structurally stable porous media that maintains a constant pore geometry throughout the flow process (Lal, 1991; Lebron and Suarez, 1992; Logsdon and Jaynes, 1993, 1996; Connolly, 1998; van Es et al., 1999; Pachepsky et al., 2001; Hodnett and Tomasella, 2002; Pachepsky and Rawls, 2003; Simunek et al., 2003; Horn, 2004; Ahuja et al., 2006; Tomer et al., 2006; Strudley et al., 2008; Roger-Estrade et al., 2009).

Soil structure is subjected to dynamic changes which, in turn, alter soil hydraulic and transport properties and its mechanical characteristics. Water storage and flow in soils depend, to a large extent, on the intricate nature of and changes in soil structure at near saturation and soil PSD. Characterization of soil structure by hydraulic properties, as well as by aggregate size distribution and stability, has been identified in numerous studies and review papers (Baver, 1937; Childs, 1940; Dexter, 1988; Le Bissonnais, 1996; Connolly, 1998; Amézketa, 1999; Kay and Angers, 2002; Green et al., 2003; Lin, 2003; Bronick and Lal, 2005; Strudley et al., 2008; Roger-Estrade et al., 2009). The relationship between soil structure and water content of the soil can be illustrated by a soil water retention curve which relates water content of the soil to the matric potential of the soil water and thus to the equivalent diameter of the water filled pores (Childs, 1940). A good correlation exists between the PSD and the size of soil aggregates and particles

(Quirk and Panabokke, 1962; Amemiya, 1965; Wu et al., 1990; Lebron et al., 2002; Guber et al., 2003). The total pore volume in soil, commonly calculated from the relation between bulk and particle density, is usually divided into textural (intra-aggregate) and structural (inter-aggregate) porosity (Hillel, 1998). Generally, it is known that the ease with which excessive water can be drained from a soil in the field is related to the inter-aggregated pores (>75 μm) or structure stability, that is, presence in the soil of stable aggregates; the effective capacity for holding available moisture for plants is higher in well-aggregated soils than in soils with poor aggregation (Childs, 1942, Kay and Angers, 2002).

The goal of soil structure stability tests is to give a reliable description and ranking of the behavior of soils when subjected to forces exerted by water, wind and management (Amezketa, 1999). The difficulty to quantify the impact of soil type and properties, conditions prevailing in the field coupled with management practices on soil structure stability, neither by empirically based nor by conceptual models, has been widely recognized (e.g., Lal, 1991; Skidmore et al., 1994; Connolly, 1998, Haynes and Naidu, 1998; Greene and Hairsine, 2004; Bronick and Lal. 2005; Ahuja et al., 2006). Consequently, measurements of the effects of soil type and clay mineralogy on hydraulic properties, crusting potential and structure stability yielded inconsistent results (e.g., Frenkel et al., 1978; Mamedov et al., 2001; Levy and Mamedov, 2002; Shainberg et al., 2003; Norton et al., 2006; Bhardwaj et al., 2007a; Reichert et al., 2009).

The primary causes for the deterioration of soil structure near the soil surface are (Or and Ghezzehei, 2002) (i) mechanical compaction by agricultural implements, (ii) surface crusting due to raindrop impact and water ponding, and (iii) subsurface rejoining of aggregates due to tensile forces of capillary water. The uncertainty clouding the effects of soil structure stability on hydraulic properties and runoff generation is associated with variety of physical and physicochemical mechanisms of soil aggregates breakdown by water (Le Bissonnais, 1996), such as (i) slaking, that is, breakdown caused by compression of entrapped air during fast wetting (Panabokke and Quirk, 1957), (ii) breakdown by differential swelling during fast wetting (Kheyrabi and Monnier, 1968), (iii) breakdown by impact of raindrops (McIntyre, 1958); and (iv) physicochemical dispersion because of osmotic stress on wetting with low electrolyte water (Emerson, 1967).

These four mechanisms may act together and thus the determination of boundary conditions may define which of the mechanisms is the most significant (Amezketa, 1999). However, they differ in the type of energy involved in aggregate disruption: for example, swelling can overcome attractive pressures in the magnitude of megapascals, while slaking and impact of raindrops can overcome attractive pressures in the range of kilopascals only (Rengasamy and Olsson,

1991; Rengasamy and Sumner, 1998). These mechanisms also differ in the size distribution of the disrupted products, and in the type of soils and soil properties affecting the mechanisms: for example, slaking may be affected by porosity and internal cohesion, while physicochemical dispersion by clay mineralogy, and the ionic composition and concentration of the soil solution (Emerson, 1967; Panabokke and Quirk, 1957; Chan and Mullins, 1994; Le Bissonnais, 1996, Norton et al., 2006).

Many different aggregate stability measurements (e.g., wet sieving, drop test technique, application of ultrasonic energy), employing diverse primary breakdown mechanisms have been used for establishing an index of soil structure. Not surprising, use of the various tests results in a wide ranking of soil structure stability, thus making the comparison of the data difficult (North, 1976; Farres, 1980; Kemper and Rosenau, 1986; Chan and Mullins, 1994; Loch, 1994; Le Bissonnais, 1996; Amézketa, 1999; Levy and Mamedov, 2002). The variety of methods and thus the obtained indices make it also complicated to link changes in those indices to soil structure and crop production relations which in turn depend on seasonal variation and soil properties (Connolly, 1998). The wet sieving is the most frequently used method, but results obtained in this technique are difficult to reproduce and hence comparison of different sample populations entails a great amount of work to obtain a significant outcome (Amézketa, 1999).

High Energy Moisture Characteristic (HEMC) Method

Among the methods used to assess soil structure and aggregate stability, yet not a very common one, is the high energy moisture characteristic (HEMC) method (Childs, 1942; Feng and Browning, 1947, Collis-George and Figueroa, 1984; Pierson and Mulla, 1989; Levy and Mamedov, 2002). In this method, the wetting process is accurately controlled, and energy of hydration and entrapped air are the main forces responsible for breaking down of aggregates. Structure and aggregate stability is inferred from changes in the PSD following wetting. The HEMC procedure has been tested in numerous studies (Table 3–1), and it has been concluded that this method is sensitive and capable of detecting even small changes in aggregate and structure stability of a range of soils from arid and humid zones (Mamedov et al., 2010).

The HEMC method was first proposed by Childs (1940), who initially compared the effects of quick wetting relative to those of slow wetting on the stability of clay soils, to evaluate draining capacity of subsoils. This approach was later modified by Collis-George and Figueroa (1984) who proposed to use a structural index (SI) as a convenient method for interpreting the experimental results for a given soil treatment. The ratio of the SI indices for two soil treatments

Table 3–1. List of papers using HEMC method for a range of soil from arid and humid zones.

No.	Authors	Purpose of using HEMC method	Location
1	Childs (1940, 1942)	Use of soil water retention, HEMC methodology, modal suction and stability of >30 clay soils	USA
2	Feng and Browning (1947)	The role of cropping system, saturation– drainage on stability of silty loam soil, Iowa	USA
3	Collis-George and Figueroa (1984)	Methodology of HEMC, structural index, examples on the effect of aggregate size, retreatments such as air drying, grinding and sieving on structural index and stability ratio of several soils	Australia
4	Levy and Miller (1997)	Aggregate stabilities of 11 GA kaolinitic soils (from loam to clay) for two water quality, relation between seal formation, erosion and aggregate stability indexes	USA
5	Crescimanno and Provenzano (1999)	The influence of sodicity on structural index (SI) of two semiarid clay soils, relation between SI and hydraulic conductivity	Italy
6	Bearden (2001)	Impact of arbuscular mycorrhizal fungi on soil structure and water retention in clay vertisol	India
7	Pierson and Mulla (1989)	Modified HEMC to measure aggregate stability of a weakly aggregated silty loam loessial soil, calculation of stability indexes and model parameters	USA
8	Pierson and Mulla (1990)	Effect of topographic landscape position (164 samples from summit, shoulder, footslope and toeslope) on aggregate stability of silty loam, Washington	USA
9	Mulla et al. (1992)	Temporal variation in silt loam aggregate stability on conventional and alternative farms at 3 slopes and seasons, Washington	USA
10	Levy and Mamedov (2002)	Modifying HEMC, accurately controlling wetting, equipment on one sheet to measure water retention characteristic; aggregate stability as a predictor for seal formation for 24 semiarid smectitic soils varying in soil texture (from sandy to clay).	Israel
11	Levy et al. (2002)	The role of water quality and soil texture and sodicity on soil bulk density and hydraulic conductivity	Israel
12	Levy et al. (2003)	Sodicity and water quality effects (distilled water vs. saline) on aggregate slaking of 56 semiarid Israeli soil widely varied in texture (from sandy to heavy clay) and sodicity (from non sodic to very high sodicity)	Israel
13	Norton et al. (2006)	The role of clay mineralogy and soil texture and tillage on soil aggregate stability from arid and humid zone soils (>20), relationship between aggregate stability and soil loss	USA, Israel, Australia, Brazil
14	Mamedov et al. (2007)	Effect of polyacrylamide molecular weight, soil texture (four soils, from loam to clay) and water quality (distilled water vs. gypsum solution) on soil aggregate and structural stability.	Israel
15	Johnson-Maynard et al. (2007)	Effect of earthworm dynamics on silt loam soil structure and aggregate stability under no-till and conventional tillage, Idaho	USA
16	Bhardwaj et al. (2007a)	The role of irrigation methods (drip vs. mist) on aggregate stability and hydraulic conductivity in semiarid clay soil	Israel
17	Bhardwaj et al. (2007b)	Effect of cross-linked polyacrylamides on water retention and structural stability in sandy soils	Israel

Table continued.

Table 3–1. Continued.

No.	Authors	Purpose of using HEMC method	Location
18	Bhardwaj et al. (2008)	The role of irrigation water quality on aridic silty clay aggregate stability, selecting irrigation water type (saline-sodic vs. treated waste)	Israel
19	Mandal et al. (2008)	The impact of saline-sodic water on aridic silty clay aggregate stability, hydraulic properties and erosion	Israel
20	De Campos et al. (2009)	Effect of redox and solution chemistry on aggregation of six upland humid Indiana soils with different texture and cultivation history (cultivated vs. forest)	USA
21	Mamedov et al. (2010)	The role of clay mineralogy, soil texture and polymer application on soil structure stability of 16 soils (loam and clay) from arid, semiarid, tropical and humid zones	USA, Israel, Brazil
22	Mamedov et al. (2002–2010 reported data)	Soil-HEMC model; effect of exchangeable Mg, clay mineralogy, tillage, crop rotation, land use, saturation-drainage, manure, sludge, humic acid, fulvic acid, etc. on structure indexes	USA, Israel, Australia, Spain, Brasil, Kenya

provides the stability ratio (SR), which is of relevance to erosion and pedogenetic studies of soil structure (Collis-George and Figueroa, 1984). Further modification of the method for weakly aggregated soils was made by Pierson and Mulla (1989), who introduced controlled slow and fast wetting treatments. The SR was then used to represent aggregate stability and was obtained from dividing the SI at fast wetting by the SI at slow wetting. Pierson and Mulla (1989) further proposed to determine the SI values from modeling the water retention curves (0 to −3.0 J kg^{-1}, macropores > 100 μm) of aggregates using a modified version of the "van Genuchten" equation (van Genuchten, 1980). All these steps, including the calculations, made the method more laborious even with experienced technician (Pierson and Mulla, 1989), but less time consuming than wet sieving (Kemper and Rosenau, 1986). Albeit not been widely used, the HEMC method provides researchers with (i) a sensitive alternative for studying stability of weakly aggregated or treated soils, (ii) detailed information on the effects different wetting treatments on the soil water retention curve or PSD, (iii) an alternative to immersion or shaking of aggregates in water, and (iv) some insight into the effects of cropping and management history on stability of weakly aggregated soils (Pierson and Mulla, 1990).

The HEMC method and calculation of the soil structure indices was further modified by Levy and Mamedov (2002), who prepared an apparatus (Fig. 3–1) that allowed a significantly easier and faster way (compared with previous studies) to measure water retention curves at the range of matric potential 0 to −5.0 J kg^{-1} (i.e., macropores > 60 μm) (Kay and Angers, 2002), under accurately controlled rate of wetting using a peristaltic pump. The modified method also

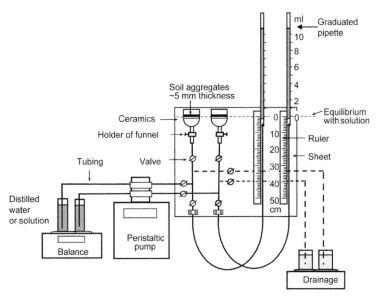

Fig. 3–1. The scheme of HEMC equipment.

enabled (i) studies of processes that are difficult to study with other methods, for example, examining successive wetting or wetting-drying treatments; (ii) use of a wide range of controlled rates of wetting (disruptive forces) and wetting solutions (distilled water, gypsum solution, tap or irrigation water, ethanol, etc.); and (iii) an easy, quick and user-friendly manner of calculating the parameters of the modified version of the "van Genuchten" equation by using the "Soil-HEMC" model (Mamedov et al., 2009).

The current review focuses on the characterization of soil structure stability indices using the HEMC method and evaluating (i) the relationship between soil SI and the parameters of the modified version of the "van Genuchten" equation (see discussion later), and (ii) the impacts of soil properties, soil management practices and extreme wetting condition on soil SI. To that end, published results (e.g., Table 3–1), that link between soil physical processes and stability indices derived by the HEMC method and their dependence on soil properties and management practices, are evaluated and discussed. The review encompasses results of PSD as determined with the HEMC method for a large body of soils (>200), collected from different climatic regions (e.g., humid, semiarid and arid: Australia, Brazil, Israel, Kenya, Spain, and USA). The soils varied in their inherent properties (e.g., clay mineralogy, texture, organic matter content), in the conditions prevailing in the soil at sampling time (salinity, type of tillage, antecedent moisture), and in several cases were also subjected to the addition of amendments (polymers, gypsum, manure, sludge) and biosolids (e.g., Levy and Miller, 1997; Levy and Mamedov, 2002; Levy et al., 2003; Norton et al., 2006; Mamedov et al., 2007; Bhardwaj et al., 2007a, 2008; De Campos et al., 2009; Mandal et al., 2008; Mamedov et al., 2010, and other reported results at various scientific meetings).

HEMC—Experimental Setup and Data Analysis

A detailed description of the most recently modified HEMC method and apparatus (Fig. 3–1) can be found in Levy and Mamedov (2002) and Mamedov et al. (2010). Briefly, in this method, macroaggregates (0.5 to 1.0 mm) are wetted from the bottom either slowly or rapidly in a controlled manner (commonly with a peristaltic pump), and then a soil water retention curve at high energies of matric potential from 0 to -5.0 J kg^{-1} (0 to 50 cm H$_2$O tension) corresponding to drainable pores of 60 to 2000 μm with small steps of 0.1–0.2 J kg^{-1} (1–2 cm), is performed. An index of structure or aggregate stability is obtained by quantifying differences between the soil water retention curves for fast and slow wetting (Fig. 3–2a and 3–2b). For a given wetting rate, the SI is defined as the ratio of volume of drainable pores (VDP) to modal suction (MS) (Collis-George and Figueroa, 1984). The MS corresponds to the matric potential (ψ, J kg^{-1}) at the peak of the specific water capacity curve ($d\theta/d\psi$) (Fig. 3–2c). The VDP is defined as the integral of the area under the specific water capacity curve and above its baseline (Fig. 3–2c). Changes in soil structure following aggregate breakdown, generally, result in the formation of a larger number of aggregates or particles of smaller sizes than the original ones. This in turn, causes the inter-aggregate or particle PSD to shift toward a greater number of smaller pores (form macro to micropores) and thus to a decrease in the VDP and to a higher value for the MS (Fig. 3–2b).

Soil structure and/or aggregate stability can be expressed in terms of SI and/ or SR. The SR is determined from the ratio of the SI indices obtained from fast and slow wetting. Generally, the SR is used to compare stability of aggregates on a relative scale of zero to one (0 < SR < 1). A SR = 0 indicates that fast wetting destroyed the aggregates to the extent that all pores that drain at the applied matric potential range no longer exist. Use of only the SI for a given treatment (e.g., fast wetted soils), or the ratio of the SI obtained from treatment A to the SI obtained from treatment B (for instance plow till vs. no-till) at a similar wetting condition is widely employed too (Collis-George and Figueroa, 1984; Crescimanno and Provenzano, 1999; Norton et al., 2006; Mamedov et al. (2007, 2010).

Traditionally, for predicting water retention data the following expression, proposed by van Genuchten (van Genuchten, 1980), is used:

$$\theta = \theta_r + (\theta_s - \theta_r) \left[1 + (\alpha \psi)^n \right]^{(1/n-1)} \tag{1}$$

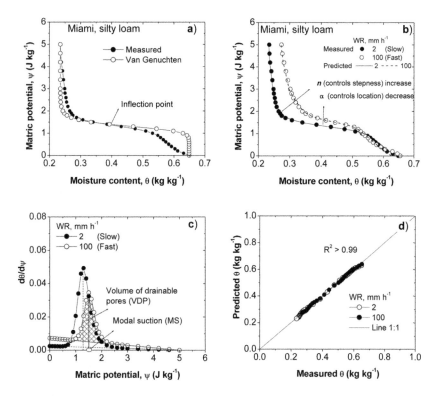

Fig. 3–2. (a) measured and fitted, by the van Genuchten model, water retention curves, (b) measured and fitted, by the Soil-HEMC model, water retention curves, (c) specific water capacity curves of a silty loam Miami (Indiana) soil for two extreme wetting rates (WR) obtained with the Soil-HEMC model; the dashed line in the specific water capacity curve represents the soil shrinkage line, and (d) the relationship between measured and predicted, by the Soil-HEMC model, water content.

where θ_r and θ_s are the residual and saturated water content (kg kg^{-1}), respectively; ψ is a matric potential (J kg^{-1}); α (m^{-1}) and n (dimensionless) represent the location of the inflection point and the steepness of the S-shaped water retention curve (Fig. 3–2b), respectively. The parameter α is related to the modal pore size, and is numerically close to the inverse of the matric potential at the inflection point. The parameter n is a function of the spread of pore-size distribution, and assumes values > 1.

When the van Genuchten equation (van Genuchten, 1980) is fitted to water retention curves obtained by the HEMC method for soil aggregates then Eq. [1] accurately predicts the inflection point, peak of the specific water capacity curve (i.e., MS), but it does not adequately predict the sloping "legs" of the curve (Fig. 3–2a) that account for pore shrinkage on desorption and thus overestimates the VDP and subsequently the SI (Table 3–2). To calculate the VDP more accurately,

Table 3–2. The various parameters and indices obtained from analyzing the water retention curve by the van Genucheten model (van Genuchten, 1980) and the modified Soil-HEMC model (Mamedov et al., 2009).

Method	WR	MS	VDP	SI	n	α	A	B	C
	mm h⁻¹	cm	g g⁻¹	cm⁻¹		cm⁻¹			
van Genuchten	2	13.04	0.414	0.032	10.55	0.076	0	0	0
	100	15.30	0.381	0.025	14.70	0.065	0	0	0
Modified	2	13.04	0.251	0.019	10.55	0.076	0.00008	–0.006	0.12
	100	15.30	0.126	0.008	14.70	0.065	0.00012	–0.010	0.23

it was suggested to model the water retention curve with the following equation (Pierson and Mulla, 1989):

$$\theta = \theta_r + (\theta_s - \theta_r) \left[1 + (\alpha\,\psi)^n\right]^{(1/n-1)} + A\psi^2 + B\psi + C \qquad [2]$$

where θ_r and θ_s (that can no longer be physically interpreted in terms of saturated and residual water contents) are the *pseudo* residual and saturated water content, respectively, and *A*, *B*, and *C* are the coefficients of the polonium added to better predict the water retention curve.

The specific water capacity curve ($d\theta/d\psi$) (Fig. 3–2c) needed for obtaining the value of VDP and MS, is computed by differentiating Eq. [2] with respect to ψ.

$$d\theta/d\psi = (\theta_s - \theta_r)\left[1 + (\alpha\,\psi)^n\right]^{(1/n-1)} (1/n - 1)(\alpha\,\psi)^n n / \left\{\psi\left[1 + (\alpha\,\psi)^n\right]\right\} + 2A\psi + B \quad [3]$$

The VDP, the area under the specific water capacity curve and above the soil shrinkage line (Fig. 3–2c), is calculated by subtracting the baseline terms for pore shrinkage ($2A\psi + B$) from Eq. [3] and analytically integrating the reminder of that equation.

Mamedov et al. (2009) have used a Soil-HEMC model containing six to eight equations comprising of the van Genuchten equation (van Genuchten, 1980) to which a polonium with two to four coefficients ($A\psi^2 + C$, $A\psi^2 + B\psi + C$, or $A\psi^3 + B\psi^2 + C\psi + D$) was added. For the parameter fittings in the model, they have used the method of least squares to find the optimal solution, which is based on solving a matrix of linear equations. This optimization procedure, generally combined with the understanding that calculated moisture content values over the entire retention curve (including θ_s and θ_r) should be close to the measured ones, allowed to choose the best parameters to be fit. Use of this model enabled an accurate fit of the water retention curves in the range of the studied matric potential (0 to –5 J kg⁻¹) for a wide variety of soils, yielding a $R^2 > 0.99$ (Fig. 3–2d) between measured and predicted θ, including the near-true θ_s and θ_r, that is, not the pseudo θ_s and θ_r obtained from the equation proposed by Pierson and

Mulla (1989). The "Soil-HEMC" model is based on optimization of parameters; it utilizes the R language (software) where the input data for the model are: ψ and the corresponding θ, and the output data are ψ, and the corresponding measured or calculated θ, calculated $d\theta/d\psi$ and the shrinkage line ($2A\psi$, $2A\psi + B$, or $3A\psi^2 + B\psi + C$). With the aid of the "Soil-HEMC" model the products of the HEMC test include: VDP and MS, and model parameters: θ_s and θ_r, α, n, A, B, C, and D, and indices of the model accuracy: R^2, LSD and RMSE.

Although both the original van Genuchten equation (van Genuchten, 1980) and the modified van Genuchten equation (Mamedov et al., 2009) for a given treatment may yield similar α, n, and MS, there is a difference in the value of VDP and subsequently in that of the SI between the two equations. It should be noted that soil VDP and SI, and the parameter α decreased and n and MS increased with the increase in wetting rate, indicating that the parameters were related to soil structure (Table 3–2, Fig. 3–2b). However using the original van Genuchten equation, may result in a >1.5 times higher VDP than that calculated by the modified equation for both wetting conditions (Table 3–2).

Relationship Between Soil Structural Index and the Model Parameters

Changes in the water retention or PSD of the surface soil layer brought about by management practices can significantly alter the amount of rain and irrigation water that infiltrates into the root zone or deeper and that is available for plant growth. Many researchers attempted to find equations (e.g., pedotransfer functions) describing the water retention curve using a simple set of measurable parameters of the soil's solid phase such as particle size distribution, bulk density, or organic matter content. These equations contain parameters which, generally, have no direct physical meaning and are mainly used as fitting parameters to match the functions to experimental points (Rawls et al., 1991; Schaap and Leij, 1998).

The shape of the soil water retention at near saturation is a dynamic soil physical property which is highly affected by soil structure (Ahuja et al., 2006). A relatively small change in the shape of the curve near saturation can significantly affect the results of the different numerical simulations (Vogel et al., 2000). Because different process control pore formation, PSD and pore continuity, the hydraulic functions are expected to differ under different tillage systems and wetting-drying conditions. These differences are most likely to occur in the larger pores, that is, at the wet end of the soil water retention curve (Logsdon et al., 1993; Ahuja et al., 1998). Moreover, even the elevated or modified inter-aggregate porosity produced after tillage and amendments application may gradually decrease by the interplay of capillary and rheological processes (e.g., van Es et al.,

1999). Therefore, the PSD and hence SI will change with time as a result of wetting and drying, solution composition, agricultural operations, and biological activity. Consequently, associated soil hydraulic and transport properties vary with time under field conditions (Leij et al., 2002).

Previously published studies that employed the HEMC method (Table 3–1) reported of a wide range of changes in PSD, and aggregate and structure stability induced by the use of soils varying in intrinsic properties and cultivation history. A study with semiarid cultivated soils (~50 soil samples) from Israel having a large spatial and temporal variation (e.g., texture, organic matter, irrigation, water quality, tillage) used two extreme wetting conditions, 2 (slow) and 100 (fast) mm h^{-1} to obtain and quantify a broad range change in the PSD and thus in the MS, VDP, and SI. It was noted that, in addition to the expected strong relationship between soil SI and VDP, and MS (Fig. 3–3), a strong relationship (exponential or power, $R^2 \geq 0.88$) exists between the SI and the parameters α and n, which control the location and steepness of the S-shape inflection of the water retention curve, as well as α/n (Fig. 3–4 and 3–5).

Soil SI increased exponentially with the increase in α and the decrease in n. The SI was a linear function of α/n, and both indices have the same unit (m^{-1}). Thus it is suggested that α/n, the ratio of parameters related to the modal pore size to the function of the spread of PSD will characterize and help to quantify soil structural condition (Fig. 3–5). Generally, for a given soil or treatment, the values of A, B (absolute value), and C increase with the increase in wetting rate, that is, the less stable the soil or the smaller the pores after wetting, the higher the values of A, B, and C. Although being a soil dependent issue, α was generally affected to a greater extent by wetting rate than A, B, and C, but the effect of wetting on n and A, B, and C was comparable. The SI also decreased exponentially (power) with an increase in the absolute values of the coefficients A (data not shown), B, and C (Fig. 3–6), however, the relationship between the SI and the A, B, or C coefficients was substantially weaker than its relationship with the α, n, and α/n parameters. The later observation could be ascribed to the fact that pore size and hence aggregate size orders vary in their response to management and environmental stresses, with larger aggregates (e.g., course, medium and fine macroporosity, pore size > 250 μm, ψ in the range of 0 to –1.2 J kg), being more susceptible to the disruptive forces than smaller aggregates (e.g., Six et al., 2004). Consequently, the shape of the legs of the measured water retention curves, that is, the shape of the polonium varied widely, and affected the relationship between SI and the coefficients (Fig. 3–6).

The aggregate size distribution and pore structure of cultivated soils can be influenced by management (e.g., the type and number of irrigation and tillage operations), but the resilience of this structure largely depends on soil genesis

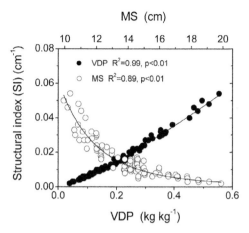

Fig. 3–3. Soil structural index (SI) as a function of the volume of drainable pores (VDP) and modal suction (MS).

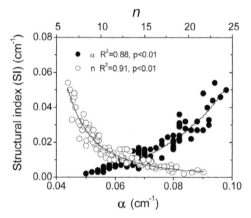

Fig. 3–4. Soil structural index (SI) as a function of model parameters α and n which control the location and steepness of the S-shape inflection of the soil water retention curve.

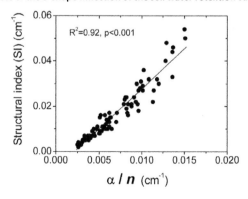

Fig. 3–5. Soil structural index (SI) as a function of the model parameter α/n.

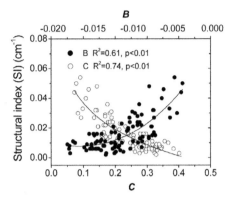

Fig. 3–6. Soil structural index (SI) as a function of the coefficients *B* and *C* of the polonium added to the water retention curve equation.

and associated properties (Kutılek, 2004; Dexter, 2004). If model parameters can be easily related to measured soil properties and or condition, then soil water retention at near saturation can be derived from easily measured field properties. The change in α and n is considered to be related to PSD and therefore to aggregate and particle size distribution (e.g., Schaap and Leij, 1998; Levy and Mamedov, 2002; Rawls and Pachepsky, 2002; Guber et al., 2004; Porebska et al., 2006; Walczak et al., 2006; Lipiec et al., 2007). Generally, under conditions of near saturation (ψ, 0 to −5.0 J kg^{-1}), the α and n parameters (i.e., parameters related to the modal pore size and to the spread of pore-size distribution, and control the location and the steepness of the S-shaped water retention curve, respectively) of the modified van Genuchten's model characterize the contribution of large (>0.25–0.5 mm, ψ, ∼0 to −1.2 J kg^{-1}) and small (<0.25–0.5 mm, ψ, ∼ −1.2 to −5.0 J kg^{-1}) aggregates/particles, respectively, to soil structure condition (see further discussion follows).

Effect of Wetting Rate and Aggregate Size

The effect of α and n, that is, the shape parameters of water retention curves, on the structural indices were variable and depended on the soil type, aggregate size, and wetting condition (Fig. 3–7 and 3–8, Tables 3–3 and 3–4).

Rate of Wetting

Data obtained for smectitic and kaolinitic loam and clay soils from semiarid Kansas and humid Hawai, USA, are used to illustrate the effects of the rate of wetting on the model parameters and structural indices (Fig. 3–7, Table 3–3). As expected, the shape of the soil water retention curve and the PSD were very sensitive to the

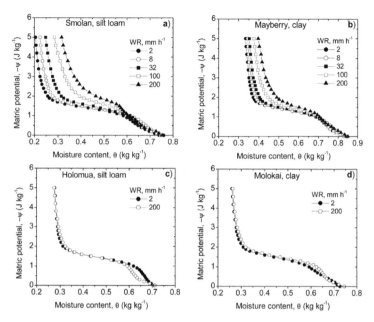

Fig. 3–7. Water retention curves of soils as affected by wetting rate (WR) and clay mineralogy. (a) Smolan silt loam and (b) Mayberry clay soils from Kansas with smectitic dominant clay mineralogy; (c) Holomua silt loam and (d) Molokai clay soils from Hawaii with kaolinitic dominant clay mineralogy.

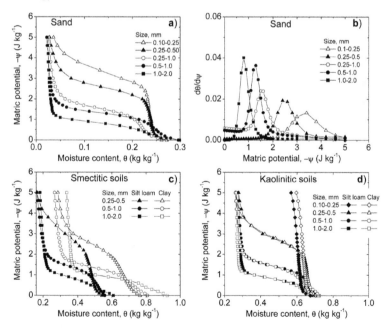

Fig. 3–8. (a) water retention and (b) specific water capacity curves of sand, and (c) water retention of smectitic soils and (d) water retention of kaolinitic soils as affected by aggregate size of slow wetted samples.

Table 3–3. Soil structural indices and model parameters as affected by wetting rate.

Soils	WR	MS	VDP	SI	n	α	α/n	A	B	C
	mm h⁻¹	cm	g g⁻	cm⁻¹		— cm⁻¹ —				
Smectitic										
Silt loam	2	13.19	0.298	0.023	9.33	0.075	0.008	0.00012	−0.009	0.19
Smolan	8	14.08	0.260	0.019	9.81	0.072	0.007	0.00012	−0.010	0.22
	32	14.48	0.243	0.017	12.41	0.070	0.006	0.00014	−0.012	0.25
	100	16.54	0.128	0.008	13.51	0.060	0.004	0.00015	−0.013	0.31
	200	19.80	0.085	0.004	13.54	0.050	0.003	0.00018	−0.013	0.32
Clay	2	12.33	0.310	0.025	8.72	0.080	0.009	0.00010	−0.008	0.15
Mayberry	8	12.51	0.297	0.024	9.21	0.079	0.009	0.00010	−0.008	0.16
	32	12.99	0.257	0.020	9.46	0.076	0.008	0.00011	−0.009	0.17
	100	13.94	0.195	0.014	10.90	0.071	0.007	0.00012	−0.010	0.23
	200	15.04	0.123	0.008	13.62	0.066	0.005	0.00016	−0.014	0.31
Kaolinitic										
Silt loam	2	13.71	0.292	0.021	9.41	0.072	0.008	0.00008	−0.006	0.12
Holomua	200	13.95	0.240	0.017	10.59	0.071	0.007	0.00008	−0.007	0.14
Clay	2	14.15	0.262	0.018	10.84	0.070	0.007	0.00009	−0.008	0.17
Molokai	200	14.36	0.203	0.014	11.28	0.069	0.006	0.00018	−0.013	0.25

Table 3–4. Structure indices and model parameters as affected by particle size for sand and soil material.

Soil or sand	Aggregate size	VDP	MS	SI	n	α	α/n
	mm	g g⁻¹	cm	cm⁻¹		——— cm⁻¹ ———	
Sand	0.10–0.25	0.185	32.93	0.005	9.52	0.030	0.003
	0.25–0.50	0.168	24.82	0.007	11.52	0.040	0.004
	0.25–1.0	0.122	15.46	0.008	10.82	0.064	0.006
	0.50–1.0	0.169	12.69	0.013	10.46	0.078	0.008
	1.0–2.0	0.194	8.02	0.024	7.10	0.122	0.017
Smectitic							
Silt loam	0.25–0.50	0.258	30.06	0.008	6.14	0.032	0.005
Nevatim	0.50–1.0	0.210	13.70	0.015	9.34	0.072	0.008
	1.0–2.0	0.225	8.50	0.026	8.62	0.115	0.014
Clay	0.25–0.50	0.236	24.56	0.010	8.88	0.040	0.005
Yagur	0.50–1.0	0.310	12.62	0.024	8.64	0.078	0.009
	1.0–2.0	0.416	8.47	0.049	7.86	0.116	0.015
Kaolinitic							
Silt loam	0.25–0.50	0.296	27.28	0.011	8.14	0.036	0.004
Holomua	0.50–1.0	0.293	13.71	0.021	9.36	0.072	0.007
	1.0–2.0	0.304	7.82	0.039	8.66	0.126	0.015
Clay	0.25–0.50	0.195	28.63	0.007	7.26	0.034	0.005
Molokai	0.50–1.0	0.205	14.14	0.015	11.12	0.070	0.006
	1.0–2.0	0.281	8.25	0.034	10.46	0.120	0.012

rate of wetting, which caused soil structural to collapse (Fig. 3–7, Table 3–3). Generally, the smectitic soils were found to be much more sensitive to slaking and to a change in PSD following changes in the rate of wetting than the kaolonitic soils. In the case of the kaolinitic soils with more stable aggregates, most of changes in the PSD occurred in a narrow range of ψ, between 0 to −1.2 J kg^{-1} (pore size > 250 μm); conversely, in the smectitic soils, with weaker aggregates, this range of ψ was much wider (−0.3 to −5.0 J kg^{-1}, pore size 60–1000 μm). Increasing the wetting rate shifted the location of water retention curve up-right (Fig. 3–7), increased θ_r, MS, and n, and decreased α, VDP and hence the SI (Table 3–3). In the smectitic soils, the effect of wetting rate remarkably depended on soil texture, being more drastic in the loam with lower clay content due to the relatively weaker structure and resistance to slaking (Fig. 3–7, Table 3–3).

The considerable effect of clay type and content on soil structure of cultivated soils could be explained by the nature of the clay minerals (Norton et al., 2006, Stern et al., 1991). Unlike the swelling smectitic soils, the mineralogical structure of non-swelling kaolinitic soils and their frequent association with iron oxides may overshadow the slaking and dispersion effect by water, leading to a small difference in PSD for the two extreme wetting rates, and hence in the model parameters and the SI in the soils of this type (Fig. 3–7, Table 3–3).

Aggregate Size

The impact of aggregate size on soil structural indices and on the model parameters was tested for pure sand (particles size of 0.10–0.25, 0.25–0.50, 0.50–1.0, and 1–2 mm), semiarid smectitic loam and clay soils from Israel and humid kaolinitic loam and clay soils from Hawaii, USA (aggregate size of 0.10–0.25, 0.25–0.50, 0.50–1.0, and 1–2 mm) (Fig. 3–8, Table 3–4). Soils with aggregates <0.25 mm and sand particles <0.10 mm did not release notable amounts of drainable water at the applied range of ψ (0 to −5 J kg^{-1}). It is concluded that for this aggregates and particles sizes the soil and the sand do not exhibit any structure and thus do not contain macropores. Therefore, 0.10- to 0.25-mm sand particles, and 0.25- to 0.50-mm soil aggregates could be considered as the higher boundary for small sand particles/soil aggregates, or the lower boundary for large particles/aggregates that contribute to soil structure and drainable pores. It should be noted that the average of these two ranges is similar to the commonly accepted size (>0.25 mm) of water-stable macroaggregates.

In the case of pure sand having stable particles with no intra-particle porosity, slaking during wetting is not an issue, therefore, increasing the size of the particles from 0.10–0.25 to 1–2 mm (i) increased the amount of inter-aggregate pores of large size and the modal pore size from ∼75 to ∼400 μm, and (ii) shifted down, vertically, the location of the water retention curves and increased the amount of drainable water at a given ψ. These two effects yielded an increase in α, α/n, and SI, and a decrease in MS (Fig. 3–8a and 3–8b, Table 3–4). Conversely, no consistent trend in the results of n was noted; increasing the size of the particles from 0.10–0.25 to 0.25–0.50 mm, increased n too, but increasing the size of particles from 0.25–0.50 to 1–2 mm decreased n (Table 3–4).

Similar to the trend noted for the sand particles, increasing the size of the soil aggregates from 0.25–0.50 mm to 1–2 mm increased α and SI, and decreased MS; the parameter n decreased with increasing the size of the aggregates from 0.50–1.0 to 1–2 mm, but increased with increasing the aggregates from 0.25–0.50 to 0.5–1.0 mm (Fig. 3–8c and 3–8d, Table 3–4). The magnitude of the changes in the model parameters and the structural indices seemed to be soil clay mineralogy and texture dependent (Fig. 3–8c and 3–8d, Table 3–4), and was probably associated with the soils' swelling–shrinkage potential and resistance to slaking (Levy and Mamedov, 2002; Levy et al., 2003).

The water retention curves of the small particles (0.10–0.25 mm) and aggregates (0.25–0.50 mm), had a shape similar to that of the van Genuchten "classic type" curve (van Genuchten, 1980) (Fig. 3–2a, 3–8a, and 3–8d). As anticipated for such curves, small variation in aggregate size distribution could notably modify the n and VDP, particularly in clay soils. The pore system of fine-textured soils is organized hierarchically, with primary pores between the textural grains and secondary pores between the aggregates, as well as macropores which are related to swelling–shrinkage processes and biological activity. Therefore the intricate effect of sand particles (0.10–0.25 mm vs. > 0.25 mm) and soil aggregates size (0.25–0.50 mm vs. > 0.50 mm) on the model parameter n, which represents the spread of aggregates PSD, could be explained by the coupled effects of soil structure and texture on the shape of the water retention curve (e.g., Guber et al., 2004; Porebska et al., 2006; Lipiec et al., 2007). Water retention by small pure sand particles (0.10–0.25 mm) and small aggregates (0.25–0.50 mm) of coarse textured soils (kaolinitic or smectitic silty loam, with a large content of sand particles) can be similar. Consequently, sand particles and soil aggregates may have a comparable PSD, and an analogous tendency changes in the n and VDP. However, in small soil aggregates water retention capacity will increase substantially with the increase in clay content. As a result, small sand particles and small soil aggregates will have different PSD (particularly for the swelling smectitic soils), and thus different tendencies in the change in n, and VDP. The observed contribution of large (>0.25–0.50 mm) and small (<0.25–0.50) mm aggregates to the PSD were related to modal pore size, α and n, and was a reflection of a intricate interaction of structure and texture and reciprocally influencing soil water retention at near saturation (Fig. 3–7 and 3–8, Tables 3–3 and 3–4).

An improved description of the PSD and its link to aggregate size and wetting rate, by using the model parameters α and n can assist in (i) the selection of management practices for obtaining the most suitable type of aggregation depending on the desired soil function, and (ii) connecting the soil pore system to water storage and transmission in cultivated soils. Introducing the SI and/or the α/n in mechanistic soil-crop models may help in better understanding and linking the effects of changes in soil structure on the soil hydraulic properties and subsequently crop yield. Existing mechanistic soil-crop models are quite complex; however, with respect to accurately modeling soil structure and expected changes thereof due to varying conditions, there is still scope for further trans-disciplinary soil physics research, and improved measurement methods (Connolly, 1998; Angulo-Jaramillo et al., 2000; Lin, 2003; Ahuja et al., 2006; Strudley et al., 2008).

The Structural Index as Affected by Soil Properties

Reports in the literature suggest that the effects of soil structure on model parameters depends also soil texture, soil clay mineralogy and organic matter content, with the latter being strongly affected by tillage (Stern et al., 1991; Hodnett and Tomasella, 2002; Bronick and Lal, 2005; Rawls and Pachepsky, 2002; Strudley et al., 2008).

Effects of soil texture and organic matter on soil SI and MS for soils with predominantly smectitic (50 mostly from the United States and Israel) or kaolinitic (32 soils, mostly from the United States) clay mineralogy, are presented in Fig. 3–9 to 3–11. In the smectitic soils, the SI increased exponentially with the increase in clay content; this phenomenon was more prominent at the slow rate of wetting, that is, under conditions of minimum aggregate breakdown by slaking. The difference in the SI between the two extreme rates of wetting increased with the increase in clay content (Fig. 3–9b), which affects bonding within aggregates, and hence structure stability of soils when wet (van Es et al., 1999; Levy et al., 2003). In the kaolinitic soils, unlike the smectitic ones, the SI and the difference in SI between the two extreme wetting rates was not affected by the soils' clay content (Fig. 3–9a) (Levy and Miller, 1997; Norton et al., 2006).

In general, for each clay mineralogy group the contribution of organic matter content (and hence tillage) on the SI was similar to the contribution of clay content, but with relatively higher magnitude for the smectitic soils under the fast wetting condition (Fig. 3–9b and 3–10b). Organic matter bonds soil particles, forms a hydrophobic coating around the aggregates and slows down aggregate wetting,

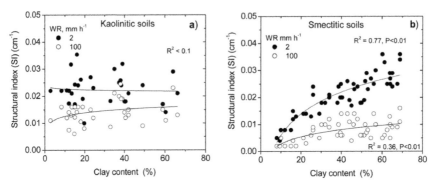

Fig. 3–9. Soil structural index (SI) as a function of clay content for two extreme wetting rates (WR): (a) kaolinitic and (b) smectitic soils.

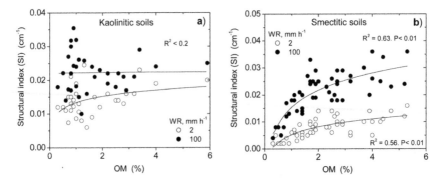

Fig. 3–10. Soil structural index (SI) as a function of organic matter (OM) for two extreme wetting rates (WR): (a) kaolinitic and (b) smectitic soils.

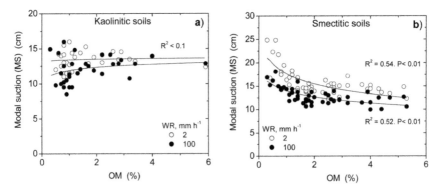

Fig. 3–11. The modal suction (MS) as a function of organic matter (OM) for two extreme wetting rates (WR): (a) kaolinitic and (b) smectitic soils.

thus reducing their sensitivity to slaking by water during wetting. In semiarid smectitic soils having low organic matter contents, the impact of clay content on the SI was more evident at slow wetting condition (Levy and Mamedov, 2002; Goldberg et al., 1988). However, the possible combined effects of clay content and organic matter may impart greater structural stability than that of clay content or organic matter alone at fast wetting. The MS values of the smectitic soils were somewhat affected by the rate of wetting rate (Fig. 3–11b) and decreased exponentially with the increase in organic matter. Conversely, in the kaolinitic soils the MS was not affected by both the organic matter and the rate of wetting (Fig. 3–11a).

The large deviation in the SI values of the soils (Fig. 3–9 and 3–10) was ascribed, in addition to differences in clay content and organic matter to large differences in the management history, soil series, climate and location of soils. Separation of the soils to types, texture, location and management may formulate more consistent data set (Strudley et al., 2008). The SI depends on the wetting conditions and could also be affected by temporal and spatial (e.g., moisture, texture) variations in pore space characteristics, that is, a structural collapse during season which is likely to take place in the late autumn and early winter to be followed by a structural recovery with the build-up of aggregates during spring and summer. For smectitic soils, the observed significant changes in the SI (e.g., changes in soil water retention, PSD and hence of the hydraulic properties), seem to depend on soil texture, and tillage treatment. These changes in SI also illustrate a dynamic characteristic of those properties, which could be of importance in soil structure and water balance modeling. The SI of kaolinitic soils also varies largely and depends on the presence of associated minerals, and soil type (Denef and Six, 2005; Levy and Miller, 1997), whereas management and wetting conditions had no clear effect on the SI in these soils. For illitic soils with mixed mineralogy, the general trend of the relationship between SI, VDP, or MS and clay content or organic matter was similar to that observed in the smectitic soils (Norton et al., 2006), but its magnitude was less pronounced (data not presented).

Understanding the effects of soil type and management interaction on soil structure in terms of the soil aggregate and structural stability, macroporosity, hydraulic conductivity, and infiltration will help the development of effective management practices to conserve soil, water, and the environment (Kay and Angers, 2002).

Characterization of Soil Structure by the HEMC Tests: Case Studies
Effect of Long-Term Tillage

Modification of soil structure by no till is a major factor in maintaining better soil aggregation compared with conventional tillage practice. The benefits of no till practice, in terms of soil structure related processes require continuity in the no till management over long periods of time, and includes enhanced soil organic carbon accumulation, since tillage operations break soil aggregates and thereby exposes soil organic matter, which was previously protected within the aggregate structure (Benjamin, 1993; Six et al., 2002).

Samples from a sandy loam soil (La Coruna, Spain), with similar properties but different tillage histories, one from a long-term tilled cornfield and one from grassland with minimum tillage (Lado et al., 2004) were subjected to the HEMC test. In comparison with the grassland soil (Fig. 3–12, Table 3–5) the long-term tillage under cornfield significantly affected soil PSD; it decreased total and residual porosity (θ_s and θ_r), water holding capacity by macropores >250 μm (i.e., macro aggregates) in the range of ψ, 0 to –1.2 J kg^{-1} (Fig. 3–12) and hence the VDP. The SI of the samples from the two different tillage regimes were comparable at slow wetting, whereas for fast wetting, the SI of the grassland soil was about fourfold higher than the SI of the cornfield soil due to the lower VDP and higher MS of former. The changes in the parameters α and n following the changes in the wetting conditions were similar to those noted in the SI and MS, respectively. For a given wetting rate, the coefficients A, B, and C were generally higher for grassland soil (Table 3–5). Under fast wetting, most of the changes in the pore size

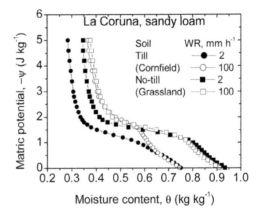

Fig. 3–12. Water retention curves of a sandy loam soil as affected by long-term tillage, La Coruna, Spain.

Table 3–5. Soil structural indices and model parameters as affected by tillage.

Tillage	WR	MS	VDP	SI	n	α	α/n	A	B	C
Crop	mm h^{-1}	cm	g g^{-1}	cm^{-1}		cm^{-1}	cm^{-1}			
Till	2	11.09	0.280	0.025	5.78	0.087	0.015	0.00009	−0.007	0.14
Cornfield	100	18.79	0.078	0.004	18.04	0.053	0.003	0.00013	−0.012	0.28
No-till	2	14.13	0.338	0.024	10.34	0.070	0.007	0.00012	−0.011	0.22
Grassland	100	15.49	0.238	0.015	12.06	0.064	0.005	0.00015	−0.013	0.28

or PSD of the more stable grassland soil occurred in the range of ψ, 0 to −1.2 J kg^{-1}, while the range of ψ for the cornfield soils was wider, −1.2 to −5.0 J kg^{-1} (Fig. 3–12).

Contrary to the cornfield soil with weak structure, the effect of fast wetting on aggregate slaking in grassland soil was less marked due to the more stable soil aggregates (associated with higher organic mater content and greater biological activity), and as a result, for both wetting conditions the location of water retention curves were relatively close each other (Fig. 3–12, Table 3–5). Moreover, the greater total porosity and water holding ability (θ_s and θ_r) in the grassland soil, were ascribed to both the inter-and intra-aggregate pore stabilization, and in turn to the stabilization of the larger and smaller aggregates which is also related to the increase in α and n. Absence of breakdown of aggregates by tillage allows the stabilization of inter-aggregate pores with a range of sizes, and smaller intra-aggregate pores by cohesion of small soil aggregates (Six et al., 2002, 2004; Strudley et al., 2008). The difference in the amount of pores with larger size between two tillage practices, substantially contributed in the grassland soils toward greater drainable pores and water storage capacity, and consequently increased the soil's saturated hydraulic conductivity and infiltration (e.g., Lado et al., 2004). Long-term tillage under corn adversely affected soil SI through deteriorating soil structure, yielding greater amounts of dispersed clay due to mechanical disruption (Lado et al., 2004) and also by reducing soil organic matter content and the microbial activity in soil (Amezketa, 1999).

Corn and Soybean Crop Rotation—No Till

Both cropping sequence and tillage system affect organic carbon accumulation, and play a major role in controlling soil structure under long-term management. It is generally acknowledged that no-till is beneficial to soil structure, and long-term rotation does not have much additional effect on soil structure. However, the quantitative prediction of the cropping outcome is usually complicated by tillage and soil interactions and also by variations in management practices, such as cover crop, etc. (Green et al., 2003; Benjamin et al., 2008, 2010).

The HEMC test was used to compare the effects of long-term (>25 yr) continues corn (*Zea mays* L.), soybean [*Glycine max* (L.) Merr.] and corn–soybean crop rotation on the structure indices and model parameters of a Throckmorton silty loam (Indiana, USA) under no-till, where crop rotation was suggested as the main factor affecting on soil structure (Norton et al., 2006). The results (Fig. 3–13, Table 3–6) of this study and literature (e.g., Benjamin, 1993; Norton et al., 2006) showed that total porosity (θ_s), water holding capacity by macropores >250 μm, and hence macro aggregates (ψ, 0 to –1.2 J kg^{-1}), and organic carbon and crop residue accumulation were in the order of contentious soybean is less than continuous corn, indicating that continuous corn exhibited much more pronounced no-till effects than either corn, soybean or corn–soybean rotations. For a given wetting condition, the θ_s, VDP, SI, and α were much higher, and θ_r, MS and n were lower for the soil under corn than under soybean. In addition to contributing to organic carbon accumulation, the residues from corn and soybeans also influenced certain soil physical properties which in turn affected soil drainage and aeration and determined the yield potential (e.g., Benjamin et al., 2008, 2010).

The absolute values of the coefficients A, B, and C depended on the wetting rate; the level of those coefficients was higher for corn and soybean at fast wetting compared with slow wetting (Table 3–6). It seems that most of the substantial changes in the PSD due to wetting condition in the soils under corn or soybean occurred in the range of ψ, –1.2 to –5.0 J kg^{-1}.

The Throckmorton silty loam (Indiana, USA) and the soil from Spain (Tables 5 and 3–6) had a comparable loam texture, mixed clay mineralogy and a similar organic matter level. For slow wetting condition, both soils, that is, under long-term no-till with continues corn (Indiana) and grassland (Spain) had comparable

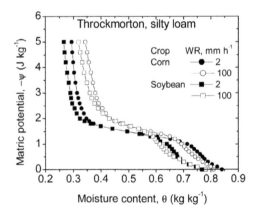

Fig. 3–13. Water retention curves of the Throckmorton silty clay soil as affected by crop rotation (corn and soybean) under long-term no-till, Lafayette, IN.

Table 3–6. Soil structural indices and model parameters as affected by crop rotation.

Crop	WR	MS	VDP	SI	n	α	α/n	A	B	C
Tillage	mm h^{-1}	cm	g g^{-1}	cm^{-1}		——cm^{-1} ——				
Corn	2	13.74	0.331	0.024	9.89	0.072	0.007	0.00011	–0.009	0.18
No-till	100	15.51	0.185	0.012	13.01	0.064	0.005	0.00014	–0.012	0.27
Bean	2	14.78	0.275	0.019	10.86	0.067	0.006	0.00010	–0.008	0.18
No-till	100	17.47	0.102	0.006	17.8	0.057	0.003	0.00015	–0.014	0.31

levels of VDP, MS, and SI and model parameters (Fig. 3–12 and 3–13, Tables 3–5 and 3–6). However, in the case of fast wetting, the VDP and SI of the grassland soil were higher by ~30% than those for soil under corn. This observation suggests that apparently even 25 yr of no-till with continuous corn could not restore the original structure and hence hydraulic properties of the soil, that is, it seems that restoration of the original structure in cultivated soils may take longer than 25 yr.

Sodicity and Salinity (Water Quality)

Sodium adsorption ratio (SAR) and electrolyte concentration (salinity) of the soil solution (irrigation or rain water) play a significant role in determining soil physical properties and the response of soil clays to slaking, dispersion and swelling. Generally a susceptibility of aggregates of semiarid soils to slaking increases with the increase in soil solution SAR (or soil exchangeable sodium percentage [ESP]), and the decrease in solution salinity (Levy et al., 2003).

The effect of sodicity and salinity on aggregate stability was determined using clay soils from Israel, namely a non sodic soil from Neve Yaar (ESP ~ 2) and moderately sodic soil from Kedma (ESP ~ 10) (Fig. 3–14, Table 3–7). Although the total porosity (θ_s) was similar for all the treatments, the decrease in wetting rate and soil sodicity, as well as increase in water salinity (changing from distilled water to tap water), decreased θ_r, MS, and n, and increased VDP, α, and SI (Fig. 3–14, Table 3–7). It transpired that sodicity and rapid wetting rate decrease the amount of macro-aggregates and thus the size of drainable pores. The absolute values of coefficients A, B, C depended on the rate of wetting and was higher for the non sodic soil at slow wetting. There were negligible differences in the MS, VDP, and SI for the slowly wetted (either with distilled or with tap water), between the two clay soils (Fig. 3–14). However, at fast wetting, the SI of the non-sodic soil, wetted with distilled water, or the SI of the sodic soil wetted with tap water was >2 times higher than the SI of the sodic soil wetted with distilled water (Table 3–7). These observations were related to the fact that for a nonsodic soil slaking by swelling has only a minor effect on aggregate dispersion, whereas, for the sodic soil both swelling and dispersion emerged as main mechanisms for

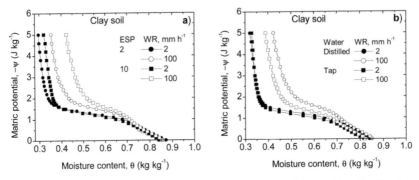

Fig. 3–14. Water retention curves of clay soils from Israel (a) nonsodic (ESP 2, Newe Yaar) and sodic (ESP 10, Kedma) soils, wetted with distilled water and (b) a sodic (ESP 10, Kedma) soil wetted with distilled and tap water.

Table 3–7. Soil structural indices and model parameters as affected by soil sodicity and water quality.

Soil sodicity	Water	WR	MS	VDP	SI	n	α	α/n	A	B	C
		mm h⁻¹	cm	g g⁻¹	cm⁻¹		—cm⁻¹—				
ESP 2	DW	2	11.83	0.361	0.031	7.95	0.083	0.010	0.00010	−0.008	0.16
		100	14.35	0.202	0.014	11.16	0.069	0.006	0.00013	−0.012	0.26
ESP 10	DW	2	10.98	0.372	0.034	6.90	0.089	0.013	0.00005	−0.004	0.09
		100	15.06	0.097	0.006	14.80	0.066	0.004	0.00017	−0.014	0.32
ESP 10	TW	2	10.27	0.369	0.036	6.63	0.095	0.014	0.00003	−0.002	0.05
		100	12.48	0.168	0.013	9.42	0.079	0.008	0.00013	−0.011	0.23

aggregates' breakdown. Moreover, these observation were useful for choosing a more suitable and available irrigation type or water quality for a given soil type (Levy et al., 2003; Bhardwaj et al., 2007a, 2008).

Concluding Comments

Characteristics of soil-structure form portray the stability and resistance to changes imposed by external forces of the soil aggregates, and controls many processes vital to agricultural practices and the environment.

The paper presents and discusses results obtained from structure and aggregate stability studies in which the HEMC method and the model (Soil-HEMC) used for analyzing the experimental results, have been employed. Aggregate stability was characterized in terms of changes in macro PSD obtained from water retention curves at near saturation (ψ, 0 to −5 J kg⁻¹). More specifically, the output results include predicted moisture content, specific water capacity, VDP and MS and model parameters. Results from >200 soil samples from humid and arid zones showed that the SI strongly depends on soil type or clay mineralogy, texture, organic matter content, and temporal conditions (e.g., wetting) prevailing in the soils. Evaluation of the behavior of the α and n parameters of the van Genuchten model produced a clearer picture of the contribution of large (>0.25–0.50 mm) and small (<0.25–0.50 mm) aggregates to the PSD and hence to soil structure stability. Furthermore, it is suggested that the use of the ratio of α/n could be applied for characterizing the effects of spatiotemporal variation on soil structural condition in soil-crop models.

Although most of the changes in soil structure take place at the inter-aggregate pore space of the upper cultivated soil layer, the current limited understandings of the complex processes that control these changes often leads many to treat the soil surface as a stable (and possibly inert) porous medium. The detailed analyses of the contribution of temporally and spatially varying soil inherent properties and extrinsic conditions to inter- and intra-aggregate porosity, SI and model parameters at near saturation, presented in the current review, may serve as a first step in improving our understanding of the processes that are involved in stabilizing soil structure, and assist in developing more efficient soil and environment management practices to achieve and maintain stable soils.

Acknowledgments

Dr. A.I. Mamedov is grateful to KSU for providing him some resources and facilities that enabled him to contribute to this work.

References

Ahuja, L.R., F. Fiedler, G.H. Dunn, J.G. Benjamin, and A. Garrison. 1998. Changes in soil water retention curves due to tillage and natural reconsolidation. Soil Sci. Soc. Am. J. 62:1228–1233. doi:10.2136/sssaj1998.03615995006200050011x.

Ahuja, L.R., L. Ma, and D.J. Timlin. 2006. Trans-disciplinary soil physics research critical to synthesis and modeling of agricultural systems. Soil Sci. Soc. Am. J. 70:311–326. doi:10.2136/sssaj2005.0207.

Amemiya, M. 1965. The influence of aggregate size on soil moisture content capillary conductivity relations. Soil Sci. Soc. Am. Proc. 29:744–748. doi:10.2136/sssaj1965.0361599500290006000 39x.

Amézketa, E. 1999. Soil aggregate stability: A review. J. Sustain. Agric. 14:83–151. doi:10.1300/J064v14n02_08.

Angers, D.A., and J. Caron. 1998. Plant-induced changes in soil structure: Processes and feedbacks. Biogeochemistry 42:55–72. doi:10.1023/A:1005944025343.

Angulo-Jaramillo, R., J.P. Vandervaere, S. Roulier, J.L. Thony, J.P. Gaudet, and M. Vauclin. 2000. Field measurement of soil surface hydraulic properties by disc and ring infiltrometers. A review and recent developments. Soil Tillage Res. 55:1–29. doi:10.1016/S0167-1987(00)00098-2.

Assouline, S. 2006. Modeling the relationship between soil bulk density and water retention curve. Vadose Zone J. 5:554–563. doi:10.2136/vzj2005.0083.

Baver, L.D. 1937. Soil characteristics influencing the movement and balance of soil moisture. Soil Sci. Soc. Am. Proc. 1:431–437. doi:10.2136/sssaj1937.03615995000100000075x.

Bearden, B.N. 2001. Influence of arbuscular mycorrhizal fungi on soil structure and soil water characteristics of vertisols. Plant Soil 229:245–258. doi:10.1023/A:1004835328943.

Benjamin, J.G. 1993. Tillage effects on near-surface soil hydraulic properties. Soil Tillage Res. 26:277–288. doi:10.1016/0167-1987(93)90001-6.

Benjamin, J.G. A.D. Halvorson, D.C. Nielsen, and M.M. Mikha. 2010. Crop management effects on crop residue production and changes in soil organic carbon in the central Great Plains. Agron. J. 102(3):990–997. doi:10.2134/agronj2009.0483.

Benjamin, J.G., M.M. Mikha, and M.F. Vigil. 2008. Organic carbon effects on soil physical and hydraulic properties in a semiarid climate. Soil Sci. Soc. Am. J. 72(5):1357–1362. doi:10.2136/sssaj2007.0389.

Bhardwaj, A.K., D. Goldstein, A. Azenkot, and G.J. Levy. 2007a. Irrigation with treated wastewater under two different irrigation methods: Effects on hydraulic conductivity of a clay soil. Geoderma 140:199–206. doi:10.1016/j.geoderma.2007.04.003.

Bhardwaj, A.K., U.K. Mandal, A. Bar-Tal, A. Gilboa, and G.J. Levy. 2008. Replacing saline–sodic irrigation water with treated wastewater: Effects on saturated hydraulic conductivity, slaking, and swelling. Irrig. Sci. 26:139–146. doi:10.1007/s00271-007-0080-1.

Bhardwaj, A.K., I. Shainberg, D. Goldstein, D.N. Warrington, and G.J. Levy. 2007b. Water retention and hydraulic conductivity of cross-linked polyacrylamides in sandy soils. Soil Sci. Soc. Am. J. 71:406–412. doi:10.2136/sssaj2006.0138.

Bronick, C.J., and R. Lal. 2005. Soil structure and management: A review. Geoderma 124:3–22. doi:10.1016/j.geoderma.2004.03.005.

Chan, K.Y., and C.E. Mullins. 1994. Slaking characteristics of some Australian and British soils. Eur. J. Soil Sci. 45:273–283. doi:10.1111/j.1365-2389.1994.tb00510.x.

Childs, E.S. 1940. The use of moisture characteristics in soil studies. Soil Sci. 50:239–250. doi:10.1097/00010694-194010000-00001.

Childs, E.S. 1942. Stability of clay soils. Soil Sci. 53:79–92. doi:10.1097/00010694-194202000-00001.

Collis-George, N. 1991. Drainage and soil structure- a review. Aust. J. Soil Res. 29:923–933. doi:10.1071/SR9910923.

Collis-George, N., and B.S. Figueroa. 1984. The use of high energy moisture characteristic to assess soil stability. Aust. J. Soil Res. 22:349–356. doi:10.1071/SR9840349.

Connolly, R.D. 1998. Modelling effects of soil structure on the water balance of soil–crop systems: A review. Soil Tillage Res. 48:1–19. doi:10.1016/S0167-1987(98)00128-7.

Crescimanno, G., and G. Provenzano. 1999. Soil shrinkage characteristic curve in clay soils: Measurement and prediction. Soil Sci. Soc. Am. J. 63:25–32. doi:10.2136/sssaj1999.03615995006300010005x.

De-Campos, A.B., A.I. Mamedov, and C. Huang. 2009. Short-term reducing condition decreases soil aggregation. Soil Sci. Soc. Am. J. 73:550–559. doi:10.2136/sssaj2007.0425.

Denef, K., and J. Six. 2005. Clay mineralogy determines the importance of biological versus abiotic processes for macroaggregate formation and stabilization. Eur. J. Soil Sci. 56:469–479. doi:10.1111/j.1365-2389.2004.00682.x.

Dexter, A.R. 1988. Advances in characterization of soil structure. Soil Tillage Res. 11:199–238. doi:10.1016/0167-1987(88)90002-5.

Dexter, A.R. 2004. Soil physical quality. Part I. Theory, effects of soil texture, density, and organic matter, and effects on root growth. Geoderma 120:201–214. doi:10.1016/j.geoderma.2003.09.004.

Emerson, W.W. 1967. A classification of soil aggregates based on their coherence in water. Aust. J. Soil Res. 5:47–57. doi:10.1071/SR9670047.

Farres, P.J. 1980. Some observations on the stability of soil aggregates to raindrop impact. Catena 7:223–231.

Feng, C.L., and G.M. Browning. 1947. Aggregate stability in relation to pore size distribution. Soil Sci. Soc. Am. J. 11:67–73. doi:10.2136/sssaj1947.036159950011000C0013x.

Frenkel, H., J.O. Goertzen, and J.D. Rhoades. 1978. Effects of clay type and content exchangeable sodium percentage and electrolyte concentration on clay dispersion and soil hydraulic conductivity. Soil Sci. Soc. Am. J. 42:32–39. doi:10.2136/sssaj1978.03615995004200010008x.

Goldberg, S., D.L. Suarez, and R.A. Glaubig. 1988. Factors affecting clay dispersion and aggregate stability of arid-zone soils. Soil Sci. 146:317–325. doi:10.1097/00010694-198811000-00004.

Green, R.T., L.R. Ahuja, and J.G. Benjamin. 2003. Advances and challenges in predicting agricultural management effects on soil hydraulic properties. Geoderma 116:3–27. doi:10.1016/S0016-7061(03)00091-0.

Greene, R.S.B., and P.B. Hairsine. 2004. Elementary processes of soil-water interaction and thresholds in soil surface dynamics: A review. Earth Surf. Process. Landf. 29:1077–1091. doi:10.1002/esp.1103.

Guber, A.X., W.J. Rawls, E.V. Shein, and Y.A. Pachepsky. 2003. Effect of soil aggregate size distribution on water retention. Soil Sci. 168:223–233.

Guber, A., Y. Pachepsky, E. Shein, and W.J. Rawls. 2004. Soil aggregates and water retention. In: Y. Pachepsky and W.J. Rawls, editors, Development of pedotransfer functions in soil hydrology. Elsevier, New York. p. 143–151.

Hamza, M.A., and W.K. Anderson. 2005. Soil compaction in cropping systems. A review of nature, causes and possible solutions. Soil Tillage Res. 82:121–145. doi:10.1016/j.still.2004.08.009.

Haynes, R.J., and R. Naidu. 1998. Influence of lime, fertilizer and manure applications on soil organic matter content and soil physical conditions: A review. Nutr. Cycl. Agroecosyst. 51:123–137. doi:10.1023/A:1009738307837.

Hillel, D. 1998. Environmental soil physics. Academic Press, San Diego, CA.

Hodnett, M.G., and J. Tomasella. 2002. Marked differences between van Genuchten soil water-retention parameters for temperate and tropical soils: A new water-retention pedo-transfer function developed for tropical soils. Geoderma 108:155–180. doi:10.1016/S0016-7061(02)00105-2.

Horn, R. 2004. Time dependence of soil mechanical properties and pore functions for arable soils. Soil Sci. Soc. Am. J. 68:1131–1137. doi:10.2136/sssaj2004.1131.

Horn, R., H. Taubner, M. Wuttke, and T. Baumgartl. 1994. Soil physical properties related to soil structure. Soil Tillage Res. 30:187–216. doi:10.1016/0167-1987(94)90005-1.

Jarvis, N.J. 2007. A review of non-equilibrium water flow and solute transport in soil macropores: Principles, controlling factors and consequences for water quality. Eur. J. Soil Sci. 58:523–546. doi:10.1111/j.1365-2389.2007.00915.x.

Johnson-Maynard, J.L., K.J. Umiker, and S.O. Guy. 2007. Earthworm dynamics and soil physical properties in the first three years of no-till management. Soil Tillage Res. 94:338–345. doi:10.1016/j.still.2006.08.011.

Karlen, D.L., E.G. Hurley, S.S. Andrews, C.A. Cambardella, D.W. Meek, M.D. Duffy, and A.P. Mallarino. 2006. Crop rotation effects on soil quality at three northern corn/soybean belt locations. Agron. J. 98:484–495. doi:10.2134/agronj2005.0098.

Kay, B.D., and D.A. Angers. 2002. Soil structure. In: A.W. Warrick, editor, Soil physics companion. CRC Press, Boca Raton, FL. p. 249–295.

Kemper, W.D., and R.C. Rosenau. 1986. Aggregate stability and size distribution. In: A. Klute, editor, Methods of soil analysis: Part 1. Physical and mineralogical methods, 2nd ed. ASA and SSSA, Madison, WI.

Kheyrabi, D., and G. Monnier. 1968. Etude experimentale de l'influence de la composition granulometrique des terres leur stabilite structurale. Ann. Agron. 19:129–152.

Kutílek, M. 2004. Soil hydraulic properties as related to soil structure. Soil Tillage Res. 79:175–184. doi:10.1016/j.still.2004.07.006.

Lado, M., A. Paz, and M. Ben-Hur. 2004. Organic matter and aggregate size interactions in infiltration, seal formation, and soil loss. Soil Sci. Soc. Am. J. 68:935–942. doi:10.2136/sssaj2004.0935.

Lal, R. 1991. Soil structure and sustainability. J. Sustain. Agric. 1:67–92. doi:10.1300/J064v01n04_06.

Le Bissonnais, Y. 1996. Aggregate stability and assessment of soil crustability and erodibility. I. Theory and methodology. Eur. J. Soil Sci. 47:425–437. doi:10.1111/j.1365-2389.1996.tb01843.x.

Lebron, I., and D.L. Suarez. 1992. Variations in soil stability within and among soil types. Soil Sci. Soc. Am. J. 56:1412–1421. doi:10.2136/sssaj1992.03615995005600050014x.

Lebron, I., D.L. Suarez, and M.G. Schaap. 2002. Soil pore size and geometry as a result of aggregate size distribution and chemical composition. Soil Sci. 167:165–172. doi:10.1097/00010694-200203000-00001.

Leij, F.J., T.A. Ghezzehei, and D. Or. 2002. Analytical models for soil pore-size distribution after tillage. Soil Sci. Soc. Am. J. 66:1104–1114. doi:10.2136/sssaj2002.1104.

Levy, G.J., and A.I. Mamedov. 2002. High-energy-moisture-characteristic aggregate stability as a predictor for seal. Soil Sci. Soc. Am. J. 66:1603–1609. doi:10.2136/sssaj2002.1603.

Levy, G.J., A.I. Mamedov, and D. Goldstein. 2003. Sodicity and water quality effects on slaking of aggregates from semi-arid soils. Soil Sci. 168:552–562. doi:10.1097/01.ss.0000085050.25696.52.

Levy, G.J., and W.P. Miller. 1997. Aggregate stabilities of some Southeastern U.S. soils. Soil Sci. Soc. Am. J. 61:1176–1182. doi:10.2136/sssaj1997.03615995006100040024x.

Levy, G.J., N. Sharshekeev, and G.L. Zhuravskaya. 2002. Water quality and sodicity effects on soil bulk density and conductivity in interrupted flow. Soil Sci. 167:692–700. doi:10.1097/00010694-200210000-00007.

Lin, H.S. 2003. Hydropedology: Bridging disciplines, scales, and data. Vadose Zone J. 2:1–11. doi:10.2136/vzj2005.0058.

Lipiec, J., R. Walczak, B. Witkowska-Walczak, A. Nosalewicz, A. Słowińska-Jurkiewicz, and C. Sławinski. 2007. The effect of aggregate size on water retention and pore structure of two silt loam soils of different genesis. Soil Tillage Res. 97:239–246. doi:10.1016/j.still.2007.10.001.

Loch, R.J. 1994. Structure breakdown on wetting. In: H.B.So et al., editor, Sealing crusting and hardsetting soils. Australian Soc. Soil Sci., Old Branch, Brisbane, Australia. p. 113–132.

Logsdon, S.D. 2002. Determination of preferential flow model parameters. Soil Sci. Soc. Am. J. 66:1095–1103. doi:10.2136/sssaj2002.1095.

Logsdon, S.D., and D.B. Jaynes. 1993. Methodology for determining hydraulic conductivity with tension infiltrometers. Soil Sci. Soc. Am. J. 57:1426–1431. doi:10.2136/sssaj1993.03615995005700060005x.

Logsdon, S.D., and D.B. Jaynes. 1996. Spatial variability of hydraulic conductivity in a cultivated field at different times. Soil Sci. Soc. Am. J. 60:703–709. doi:10.2136/sssaj1996.03615995006000030003x.

Logsdon, S.D., J. Jordahl, and D.L. Karlen. 1993. Tillage and crop effects on ponded and tension infiltration rates. Soil Tillage Res. 28:179–189. doi:10.1016/0167-1987(93)90025-K.

Malone, R.W., S. Logsdon, M.J. Shipitalo, J. Weatherington, L.R. Ahuja, and L. Ma. 2003. Tillage effects on macroporosity and herbicide transport in percolate. Geoderma 116:191–216. doi:10.1016/S0016-7061(03)00101-0.

Mamedov, A.I., S. Beckmann, C. Huang, and G.J. Levy. 2007. Aggregate stability as affected by polyacrylamide molecular weight, soil texture and water quality. Soil Sci. Soc. Am. J. 71:1909–1918. doi:10.2136/sssaj2007.0096.

Mamedov, A.I., G.J. Levy, F.A. Aliev, L.E. Wagner, and F. Fox. 2009. High energy moisture characteristics: Linking between soil physical processes and structure stability. 2009 International Annual Meetings, 1–5 Nov. 2009. ASA-CSSA-SSSA Pittsburgh, PA.

Mamedov, A.I., I. Shainberg, and G.J. Levy. 2001. Irrigation with effluent: Effect of prewetting rate and clay content on runoff and soil loss. J. Environ. Qual. 30:2149–2156. doi:10.2134/jeq2001.2149.

Mamedov, A.I., L.E. Wagner, C. Huang, L.D. Norton, and G.J. Levy. 2010. Polyacrylamide effects on aggregate and structure stability of soils with different cay mineralogy, 7. Soil Sci. Soc. Am. J. 74:1720–1732. doi:10.2136/sssaj2009.0279.

Mandal, U.K., A.K. Bhardwaj, D.N. Warrington, D. Goldstein, A. Bar Tal, and G.J. Levy. 2008. Changes in soil hydraulic conductivity, runoff, and soil loss due to irrigation with different types of saline–sodic water. Geoderma 144:509–516. doi:10.1016/j.geoderma.2008.01.005.

McIntyre, D.S. 1958. Permeability measurement of soil crusts formed by raindrop impact. Soil Sci. 85:185–189.

Mulla, D.J., L.M. Huyck, and J.P. Reganold. 1992. Temporal variation in aggregate stability on conventional and alternative farms. Soil Sci. Soc. Am. J. 56:1620–1624. doi:10.2136/sssaj1992.03615995005600050047x.

North, P.F. 1976. Towards an absolute measurement of soil structural stability using ultrasound. J. Soil Sci. 27:451–459. doi:10.1111/j.1365-2389.1976.tb02014.x.

Norton, L.D., A.I. Mamedov, G.J. Levy, and C. Huang. 2006. Soil aggregate stability as affected by long-term tillage and clay mineralogy. Adv. Geoecol. 38:422–429.

Or, D., and T.A. Ghezzehei. 2002. Modeling post-tillage soil structural dynamics: A review. Soil Tillage Res. 64:41–59. doi:10.1016/S0167-1987(01)00256-2.

Pachepsky, Y.A., and W.J. Rawls. 2003. Soil structure and pedotransfer functions. Eur. J. Soil Sci. 54:443–451. doi:10.1046/j.1365-2389.2003.00485.x.

Pachepsky, Y.A., W.J. Rawls, and D. Gimenez. 2001. Comparison of soil water retention at field and laboratory scales. Soil Sci. Soc. Am. J. 65:460–462. doi:10.2136/sssaj2001.652460x.

Panabokke, C.R., and J.P. Quirk. 1957. Effect of initial water content on stability of soil aggregates in water. Soil Sci. 83:185–195. doi:10.1097/00010694-195703000-00003.

Peigne, J., B.C. Ball, J. Roger-Estrade, and C. David. 2007. Is conservation tillage suitable for organic farming? A review. Soil Use Manage. 23:129–144. doi:10.1111/j.1475-2743.2006.00082.x.

Pierson, F.B., and D.J. Mulla. 1989. An improved method for measuring aggregate stability of a weakly aggregated loessial soil. Soil Sci. Soc. Am. J. 53:1825–1831. doi:10.2136/sssaj1989.03615995005300060035x.

Pierson, F.B., and D.J. Mulla. 1990. Aggregate stability in the Palouse region of Washington: Effect of landscape position. Soil Sci. Soc. Am. J. 54:1407–1412. doi:10.2136/sssaj1990.03615995005400050033x.

Porebska, D., C. Slawinski, K. Lamorski, and R.T. Walczak. 2006. Relationship between van Genuchten's parameters of the retention curve equation and physical properties of soil solid phase. Int. Agrophys. 20:153–159.

Quirk, J.P., and C.R. Panabokke. 1962. Pore volume-size distribution and swelling of natural soil aggregates. Eur. J. Soil Sci. 13:71–81. doi:10.1111/j.1365-2389.1962.tb00683.x.

Rawls, W.J., T.J. Gish, and D.L. Brakensiek. 1991. Estimation of soil water retention from soil properties—A review. Adv. Soil Sci. 16:213–235. doi:10.1007/978-1-4612-3144-8_5.

Rawls, W.J., and Ya.A. Pachepsky. 2002. Soil consistence and structure as predictors of water retention. Soil Sci. Soc. Am. J. 66:1115–1126. doi:10.2136/sssaj2002.1115.

Reichert, J.M., L.D. Norton, N. Favaretto, C. Huang, and E. Blume. 2009. Settling velocity, aggregate stability, and interrill erodibility of soils varying in clay mineralogy. Soil Sci. Soc. Am. J. 73:1369–1377. doi:10.2136/sssaj2007.0067.

Rengasamy, P., and K.A. Olsson. 1991. Sodicity and soil structure. Aust. J. Soil Res. 29:65–76.

Rengasamy, P., and M.E. Sumner. 1998. Processes involved in sodic behavior. In: M.E. Sumner and R. Naidu, editors, Sodic soils. Oxford Univ. Press. p. 35–50.

Roger-Estrade, J., G. Richard, A.R. Dexter, H. Boizard, S. de Tourdonnet, M. Bertrand, and J. Caneill. 2009. Integration of soil structure variations with time and space into models for crop management. A review. Agron. Sustain. Dev. 29:135–142. doi:10.1051/agro:2008052.

Schaap, M.G., and F.J. Leij. 1998. Database-related accuracy and uncertainty of pedotransfer functions. Soil Sci. 163:765–779. doi:10.1097/00010694-199810000-00001.

Shainberg, I., A.I. Mamedov, and G.J. Levy. 2003. Role of wetting rate and rain energy on seal formation and erosion. Soil Sci. 168:54–62. doi:10.1097/00010694-200301000-00007.

Shouse, P.J., and B.P. Mohanty. 1998. Scaling of near-saturated hydraulic conductivity measured using disc infiltrometers. Water Resour. Res. 34:1195–1205. doi:10.1029/98WR00318.

Simunek, J., N.J. Jarvis, M.T. van Genuchten, and A. Gardenas. 2003. Review and comparison of models for describing non-equilibrium and preferential flow and transport in the vadose zone. J. Hydrol. 272:14–35. doi:10.1016/S0022-1694(02)00252-4.

Six, J., H. Bossuyt, S. de Gryze, and K. Denef. 2004. A history of research on the link between (micro) aggregates, soil biota, and soil organic matter dynamics. Soil Tillage Res. 79:7–31. doi:10.1016/j.still.2004.03.008.

Six, J., C. Feller, K. Denef, S.M. Ogle, J.C. de Moraes Sa, and A. Albrecht. 2002. Soil organic matter, biota and aggregation in temperate and tropical soils—Effects of no-tillage. Agronomie 22:755–775. doi:10.1051/agro:2002043.

Skidmore, E.L., L.J. Hagen, D.V. Armbrust, A.A. Durar, D.W. Fryrear, K.N. Potter, L.E. Wagner, and T.M. Zobeck. 1994. Methods for investigating basic processes and conditions affecting wind erosion. In: R. Lal, editor, Soil erosion. Research methods. St. Lucie Press, Delray Beach, FL. p. 295–330.

Stern, R., M. Ben-Hur, and I. Shainberg. 1991. Clay mineralogy effect on rain infiltration, seal formation and soil losses. Soil Sci. 152:455–462. doi:10.1097/00010694-199112000-00008.

Strudley, M.W., T.R. Green, and J.C. Ascough. 2008. Tillage effects on soil hydraulic properties in space and time: State of the science. Soil Tillage Res. 99:4–48. doi:10.1016/j.still.2008.01.007.

Sumner, M.E., and B.A. Stewart. 1992. Soil crusting: Chemical and physical processes. Advanced Soil Sci. Lewis, Boca Raton, FL.

van Es, H.M. 1993. Evaluation of temporal, spatial, and tillage-induced variability for parameterization of soil infiltration. Geoderma 60:187–199. doi:10.1016/0016-7061(93)90026-H.

van Es, H.M., C.B. Ogden, R.L. Hill, R.R. Schindelbeck, and T. Tsegaye. 1999. Integrated assessment of space, time, and management-related variability of soil hydraulic properties. Soil Sci. Soc. Am. J. 63:1599–1608. doi:10.2136/sssaj1999.6361599x.

van Genuchten, M.Th. 1980. A closed form equation for predicting the hydraulic conductivity of unsaturated soils. Soil Sci. Soc. Am. J. 44:892–898. doi:10.2136/sssaj1980.03615995004400 050002x.

Vogel, H.J., and K. Roth. 1998. A new approach for determining effective soil hydraulic functions. Eur. J. Soil Sci. 49:547–556. doi:10.1046/j.1365-2389.1998.4940547.x.

Vogel, T., M.Th. van Genuchten, and M. Cislerova. 2000. Effect of the shape of the soil hydraulic functions near saturation on variably-saturated flow predictions. Adv. Water Resour. 24:133–144. doi:10.1016/S0309-1708(00)00037-3.

Tomer, M.D., C.A. Cambardella, D.E. James, and T.B. Moorman. 2006. Surface-soil properties and water contents across two watersheds with contrasting tillage histories. Soil Sci. Soc. Am. J. 70:620–630. doi:10.2136/sssaj2004.0355.

Walczak, R.T., F. Moreno, C. Sławinski, E. Fernandez, and J.L. Arrue. 2006. Modeling of soil water retention curve using soil solid phase parameters. J. Hydrol. 329:527–533. doi:10.1016/j.jhydrol.2006.03.005.

Wu, L., J.A. Vomocil, and S.W. Childs. 1990. Pore size, particle size, aggregate size and water retention. Soil Sci. Soc. Am. J. 54:952–956. doi:10.2136/sssaj1990.03615995005400040002x.

Chapter 4

Applications of Complex Network Models to Describe Soil Porous Systems

Dean Korošak and Sacha Jon Mooney

Abstract

Network science has been successful in describing and exploring the structure of complex systems over the last decade. However, only recently have the concepts of network theory been applied to complex arrangement of pores in soils. Here, we present and compare two complex network models of soil pore structure after first giving a short overview of the network theory. We show that both models point to a scale-free network organization of soil pore space and can be related to the network properties of soil structure given by the image analysis of images of soil structure obtained from scanning undisturbed soil cores using X-ray computed tomography. Furthermore, we show that binary images generated by network growth mechanism simulating soil structure statistically resemble images collected from the undisturbed soil samples. Using the relationship between the network structure and the spatial correlation properties of the soil pore structure, we discuss the possible implications of network theory applications to understand soil biophysical functioning such as self-organization of soil habitats.

Abbreviations: CT, computed tomography; EN, evolving network model; SN, static network model.

D. Korošak, Applied Physics, Faculty of Civil Engineering, Univ. of Maribor, Smetanova ulica 17, Maribor SI-2000, Slovenia, and Faculty of Medicine, Institute of Physiology, Univ. of Maribor, Slomškov trg 15, Maribor SI-2000, Slovenia (dean.korosak@uni-mb.si). S.J. Mooney, School of Biosciences, Univ. of Nottingham, The Gateway Building, Sutton Bonington Campus, Leicestershire, LE12 5RD, UK (sacha.mooney@nottingham.ac.uk).

doi:10.2134/advagricsystmodel3.c4

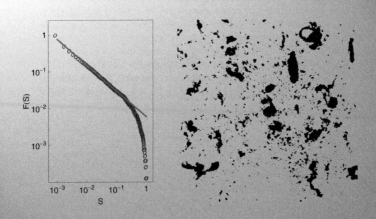

With increasing anthropogenic demands on soils (Lal, 2007), there is a growing need to seek for novel approaches to understand the relationships between soil porous structure and soil function. The spatial organization of soil in particular has a profound impact on its function, so the early emergence of network models (Fatt, 1956; Adler and Brenner, 1984; Adler, 1985) attempting to describe soil pore structure is not surprising. Numerous approaches to describe and model the complex structure of soils have been developed including fractal theory (e.g., Posadas et al., 2003; Perfect et al., 2009), Boolean random sets, cellular automata and network models (Horgan and Ball, 1994; Vogel and Roth, 2001; Prosperini and Perugini, 2007). The increasing availability of high resolution three-dimensional imaging techniques of porous media, such as X-ray computed tomography (CT) (see Chapter 6, Peth et al., 2013, this volume) has lead to the advent of powerful computational techniques to extract the pore space topology from images of soil structure (Ioannidis and Chatzis, 2000; Liang et al., 2000; Papadopoulos et al., 2009) and advancing the development of pore network modeling. Pore network models are generally based on attempts to idealize pore topology into a hydraulically similar but much simpler geometrical network often coupled with percolation theory (Berkowitz and Ewing, 1998; Hunt and Ewing, 2009) using cubic, fixed grid lattices (Vogel and Roth, 2001) with a defined connectivity function (Vogel, 1997). While this approach gives the essential physics of the random porous media, it importantly neglects the spatial correlations and the long-range heterogeneities in porous media that have a multifractal structure (Stanley and Meakin, 1988; Posadas et al., 2003; Dathe et al., 2006; Tarquis et al., 2009).

To capture both short and long-range effects in pore network models, a new approach for considering the porous architecture of soils using complex random networks has been recently suggested (Santiago et al., 2008; Mooney and Korošak, 2009; Cardenas et al., 2010). Over the last decade, network theory has become a much discussed cross-disciplinary research field contributing to social, biological and information sciences (Watts and Strogatz, 1998; Albert and Barabasi, 2002; Newman, 2003; Barabasi, 2009), demonstrating that quite diverse systems might share a similar topological organization. However, the geographical effects or spatial embedding that significantly constrains network topology and plays an important role in many "real world" systems, including soils, has not yet received the wide attention experienced in other disciplines (Yook et al., 2002; Masuda et al., 2005; Morita, 2006; Bianconi et al., 2009).

To demonstrate and further develop the application of complex networks for use in soil science and in particular soil biophysics, we here demonstrate and compare two complex network models of soil pore structure and show how they capture the structural properties of the soil porous architecture. We also show how network representations of soil pore structure are constructed using binary images of soils obtained from undisturbed soil cores scanned using X-ray computed tomography, we analyze the large-scale organization of networks before discussing the possible implications of using network theory with regard to theories concerning the self-organization of soils (Young and Crawford, 2004).

Key Concepts in Complex Network Theory

Here we give only a brief overview of the basic ideas in complex random network theory. There are several excellent review papers (Albert and Barabasi, 2002; Newman, 2003; Boccaletti et al., 2006) that describe the science of complex networks in detail. Essentially random networks are statistical ensembles of graphs where the particular graph has a specific probability of realization. Networks or graphs are rather simple objects consisting of nodes (or vertices) connected by links (or edges). Networks can be directed, where the order of the nodes i and j connected with a link (i,j) is important (the link points from node i to node j), or undirected where the order of connected nodes in unimportant. Furthermore, links can have different weights (weighted network) or each link in the network can be equally important (unweighted network). A network (unweighted and undirected) with N nodes can be completely described by an $N \times N$ adjacency matrix where the matrix element a_{ij} equals 1 if there is a link between nodes i and j or 0 otherwise. The degree of the node, k_i, is the number of links attached to node i and is given by the sum

$$k_i = \sum_j a_{ij} \qquad [1]$$

The distribution of node degrees is often used to characterize the basic topological structure of the network. The degree distribution $P(k)$ gives the probability that a randomly chosen node has k links. The first moment of $P(k)$ is the mean degree of the network $\langle k \rangle$ given by

$$\langle k \rangle = \sum k' P(k') \qquad [2]$$

In many real world networks, a particular network organization is discovered which is characterized by a power-law degree distribution, namely

$$P(k) \propto k^{-\gamma} \qquad [3]$$

with $\gamma \in [2,3]$. Such networks are termed scale-free networks due to the property of power-laws that retain the same functional form under the change of scale. Other common networks include random networks [also called Erdos-Renyi networks (Boccaletti et al., 2006)] where each couple of N nodes are connected with a probability p, resulting in a mean degree given by $\langle k \rangle = p(N - 1)$ and Poissonian degree distribution, and small-world networks realized as lattices with additional random long-range connections.

Real world systems with fat-tailed degree distributions are sometimes difficult to analyze due to rather strong noise. In such cases it is better to consider the cumulative degree distribution, namely:

$$F(k) = \sum_{k'=k} P(k') \qquad [4]$$

where statistical fluctuation are usually smoothed out and the scaling coefficient of the degree distribution is obtained from the scaling exponent of the cumulative distribution γ' as $\gamma = 1 + \gamma'$. Additionally, many real world networks are correlated meaning the probability of a node with degree k to link to a node with degree k' depends on degree k. In other words, the correlation properties of the network are expressed with a conditional probability $P(k'|k)$. The computation of the average degree of nearest neighbors with degree k is given by

$$k_{nn}(k) = \sum_{k'} k'P(k'|k) \qquad [5]$$

This is then used to classify networks into the assortative, if $k_{nn}(k)$ is an increasing function of k and disassortative if $k_{nn}(k)$ is a decreasing function of k. In assortative networks the nodes with similar degrees tend to connect, while in disassortative the nodes with low degrees will tend to connect to well connected nodes.

Describing Soil Pore Topology Using Complex Networks

The rationale for using network theory to describe the porous architecture of soil is crucially not to obtain a detailed geometrical network of pore sizes, pore throats and connectivity similar to the ones that can now be rapidly obtained through skeletonization of high-resolution three-dimensional images derived from X-ray CT but rather to provide a new way to explore and understand the complexity/ heterogeneity of soil structure through the study of large-scale organization of pore structure. In this approach we examine the statistical distributions of linkage and the range of pore sizes in the whole soil porous complex network and their relation to each other. From this we can probe and further explore the relationship with soil function, for example, what is the impact on the behavior of given microbial communities of connectivity pores and pore throat size? Here the network links do not correspond to true pore throats between the pores, but rather reflect the probability for the two pores to be connected.

Static and Evolving Complex Network Models of Soil Pore Structure

Two approaches to build complex network models of soil pore organization have recently been developed: threshold network (Mooney and Korošak, 2009) and the heterogeneous preferential attachment network(Santiago et al., 2008; Cardenas et al., 2010). Both methods use geometrical information on pore sizes and their relative spatial positions as input data obtained from the analysis of soil structure images (e.g., X-ray CT scans; Mooney and Korošak, 2009), but use different methods to construct the complex pore networks.

If we consider a set of N pores representing the nodes of the network that represent the centers of the pores and the links between nodes are derived from the probability that depends on the unique properties of the pores, in this case pore size area, and their spatial positions. We then let the nodes be distributed in D-dimensional space with the density $\rho(\mathbf{r})$ and to each node we assign a state, s, (the pore size) which describes the internal properties of the node. The node states are then distributed by a probability distribution $P(s)$ (i.e., the pore size distribution). Using a threshold network model such as that applied in Mooney and Korošak (2009), the two nodes with states s_i, s_j (pores with sizes s_i, s_j) said to be connected if

$$\frac{s_i s_j}{d_{ij}^m} > \varepsilon \qquad\qquad\qquad [6]$$

where the threshold value $\varepsilon > 0$ controls the number of links in the network, if the threshold value is set to 0 we derive with a completed network. $d_{ij} = |\mathbf{r}_i - \mathbf{r}_j|$ is the distance between pores i and j, where as the parameter m measures the importance of the distance between nodes. This procedure results in a static network of N nodes connected by M links with the mean degree of the network given by $\langle k \rangle = 2M/N$.

In the heterogeneous preferential attachment approach (Santiago et al., 2008; Cardenas et al., 2010), the complex pore network grows sequentially by adding new pores to the network and linking them to the existing ones. The probability that a connection takes place with a the new node j, randomly introduced at point

$\mathbf{r}_j$ with a state s_j drawn from the distribution $P(s)$, is proportional to the product of the degree of the existing node k_i and the affinity function $\sigma(i, j)$

$$\pi_{ij} \propto k_i \sigma(i, j) \tag{7}$$

The affinity function depends on the state of the node and the distance to the existing node:

$$\sigma(i, j) = \frac{s_j^b}{d_{ij}^m} \tag{8}$$

Here the parameter b measures the importance of the pore size in the attachment process. The growth of the network is stopped when the desired number of nodes N is reached.

Both methods of network construction described above, the static network model (SN) and the evolving network model (EN), in which the network heterogeneity can be tuned by changing the model parameters b and m, predict topologically similar networks of soil pore structures revealing the apparent scale-free topology of soils with a power law degree distribution $P(k) \propto k^{-\gamma}$. The EN model exhibits a multiscaling behavior (Bianconi and Barabasi, 2001; Santiago et al., 2008) with the degree distribution scaling exponent given by

$$\gamma = 1 + \frac{2}{w} \tag{9}$$

where w is the mean-field normalized average affinity function (Santiago et al., 2008; Cardenas et al., 2010). It measures the state of the given node relative to the average node state in the network. The states with w larger than the average ($w > 1$) will cause the existence of hubs, that is, nodes with a very large number of links in the network, and a slower decay of the degree distribution. The scaling exponent of the average nodes will be $\gamma = 3$. The multiscaling exponent in Eq. [9] depends on the choice of the affinity function and the distribution of the node states $P(s)$ or the pore size distribution function. Surprisingly, the pore size distributions obtained from a set of contrasting soils (predominately different textures ranging from sands to clays) all follow a similar scaling law $P(s) \propto s^{-\alpha}$ (Mooney and Korošak, 2009), with the scaling exponents in the interval $1 < \alpha < 2$. In Fig. 4–1 we illustrate an example of the two-dimensional soil structure image taken from an X-ray CT system and the corresponding normalized cumulative pore size distribution $F(s) = \int P(s')ds'$ extracted by image analysis of the soil pore space showing scaling behavior with $\alpha \approx 1.6$. With the power-law pore size distribution $P(s) \propto s^{-\alpha}$ the SN model predicts the scale-free organization of soil pore structure with the degree distribution $P(k) \propto k^{-\gamma}$. The scaling exponent depends

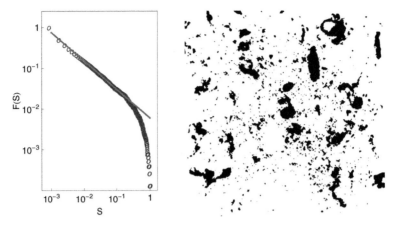

Fig. 4–1. (left) Cumulative pore size distribution determined from an X-ray computed tomography image (right). NB. Pore space is black. The cumulative distribution of the pore sizes is given by $F(S) = \int_S^\infty P(x)dx$; $F(S)$ and S are rescaled with the maximum values. The solid line indicates scaling of the pore size distribution with the scaling exponent α = 1.6 of the pore size distribution $P(S) \propto S^{-\alpha}$. The cutoff of power-law scaling (linear regime in the log-log plot) is due to the finite size of the system.

on the parameters of the model (α, m) and the embedding dimension D, which is shown by Masuda et al. (2005) as

$$\gamma = 1 + \frac{m(\alpha - 1)}{D} \tag{10}$$

Figure 4–2 shows the network development as the threshold parameter ε of the SN model is lowered to allow more and more links to appear. The formation of the network is overlain on the two-dimensional soil image for clarity. These connections will have a profound influence over the soil biophysical behavior in terms of gaseous exchange and water storage functioning. In Fig. 4–3 and 4–4 we show the soil pore networks obtained by both methods from the same set of data derived from soil image shown in Fig. 4–1 and with equal model parameters: the average number of links and $m \approx 1$. The area of particular node in Fig. 4–3 and 4–4 is proportional to its degree. The networks are presented in abstract space so the geometric positions of the nodes here do not correspond to the actual positions of pores on two-dimensional soil image as it is the case for the network in Fig. 4–2. The networks produced by both models show similar topology where the well connected nodes (larger dots) are not directly linked but are preferably linked through nodes with small degrees (smaller dots). This type of network structure indicates that networks from both models are correlated (degree correlations are considered in more detail later in the text). The large-scale of the network structure is reflected in the network degree distribution. We

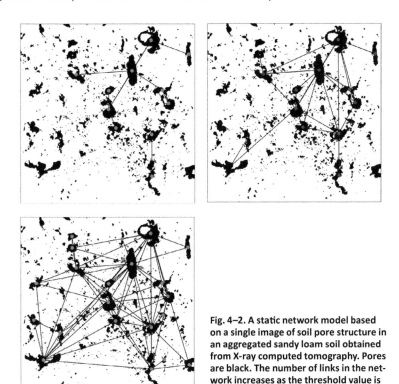

Fig. 4–2. A static network model based on a single image of soil pore structure in an aggregated sandy loam soil obtained from X-ray computed tomography. Pores are black. The number of links in the network increases as the threshold value is lowered (from left to right).

have studied the influence of the model parameter m and observed that the network structure changes with increasing value of m from scale-free to a random geometric one. In Fig. 4–5 we display the corresponding cumulative degree distributions $F(k) = \int P(k')\mathrm{d}k'$ for both models for $m \approx 1$ and $m \gg 1$. It is clear that both models produce pore networks with heterogeneous, scale-free organization for low values of the parameter m (the spatial positions of the nodes play less significant role) with $\gamma = 2$ and more compact, geometrical random pore network for large values of m with Poisson degree distribution.

Network Models and Soil Image Analysis

To build the SN and EN soil network models, pore specific data; for example, pore size and spatial distribution from image analysis of the soil structure images are required, so it is interesting to see how the structural properties of soil are subsequently reflected in the network structure. Here, we consider the correlation and (multi)fractal properties of the soil pore structure as determined from the soil structure images. The correlation properties of the statistical heterogeneous

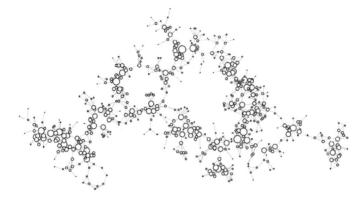

Fig. 4–3. Network representation of soil pore structure obtained by static network model. The area of nodes correspond to the node degree or the number of links attached.

Fig. 4–4. Network representation of soil pore structure obtained by growing network model. The area of nodes correspond to the node degree or the number of links attached.

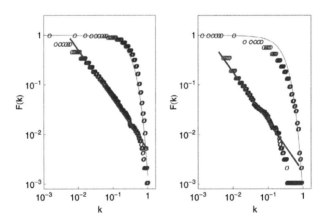

Fig. 4–5. Cumulative degree distributions of static (left) and evolving (right) network models for small and large values of the parameter *m*. The cumulative degree distribution is given by $F(S) = \int_{k}^{\infty} P(x)dx$; *F(k)* and *k* are in the figure rescaled with the maximum values The scaling exponent of the scale-free degree distribution obtained for small m (straight line in plots) is $\gamma = 2$. For large values of m the random geometrical network is obtained with the Poissonian degree distribution (curved line in plots).

porous medium can be captured by a two-point correlation function. A two-point correlation function $S_2(\mathbf{r}_1, \mathbf{r}_2)$ is given as an average of the characteristic function h:

$$S_2(\mathbf{r}_1, \mathbf{r}_2) = \langle h(\mathbf{r}_1)h(\mathbf{r}_2) \rangle \qquad [11]$$

where the characteristic function of the pore space is defined as:

$$h(\mathbf{r}) = \begin{cases} 1, & \text{if } \mathbf{r} \text{ at pore} \\ 0, & \text{otherwise} \end{cases} \qquad [12]$$

The porosity (ϕ) is related to a two-point correlation function through $\phi = S_2(0)$, and if there is no long range order, the limit $\lim_{r \to \infty} S_2(r) = \varphi^2$ is also valid (Yeong and Torquato, 1998a; Yeong and Torquato, 1998b). The correlation function $S_2(\mathbf{r}_1, \mathbf{r}_2)$ in isotropic systems depends only on the distance $r = |\mathbf{r}_1, \mathbf{r}_2|$, and the normalized two-point correlation function $g(r)$ which is usually defined as:

$$g(r) = \frac{S_2(r) - \varphi^2}{\varphi(1 - \varphi)} \qquad [13]$$

so that $g(0) = 1$ and $\lim_{r \to \infty} g(r) = 0$. The computed two-point correlation function and the normalized correlation function (averaged over two perpendicular directions) for an image of two-dimensional soil structure are shown in Fig. 4–6. The correlation function in Fig. 4–6 can be described with a power-law correlation function (Freltoft et al., 1986) $g(r) \propto r^{-(D-D_f)} \exp(-r/\varsigma)$, where D, D_f, ς are the embedding dimension, the pore fractal dimension and the upper correlation cut-off length. For the analyzed image, the fractal dimension was found to be $D_f \approx 1.7$ This is within the range of values frequently obtained from multifractal analysis of soils (Perrier et al., 2006; Perfect et al., 2009), giving us $D_f(q)$ of the same two-dimensional image, which yielded $D_f(q = 0) = 1.74$ as shown in Fig. 4–7 also the corresponding singularity spectrum $f(\alpha)$ (Halsey et al., 1986) is also presented.

The probability that two nodes at distance r are connected is considered in the SN model, given by $q(r) \propto r^{-m(\alpha-1)}$ (Masuda et al., 2005). The correlation function suggests that the probability to find a pore r away from the chosen pore is again a power-law (up to a cutoff length) with the scaling exponent $D_f - D$. Therefore, we expect the probability that the point at the distance r from a chosen pore is also a pore and that the two pores are connected in the network sense to fall off with the distance as the power-law $r^{-\delta}$ with the scaling exponent $\delta = D - D_f + m(\alpha - 1)$. Adding long-range links with probability $r^{-\delta}$ between nodes in a D-dimensional lattice resulted in a small-world network for $\delta < 2D$ (Petermann and De Los Rios, 2006) with an optimal structure for navigation at $\delta = D$ (Kleinberg, 2000). It has been suggested by Kleinberg (2000) and Petermann and De Los Rios (2006) that

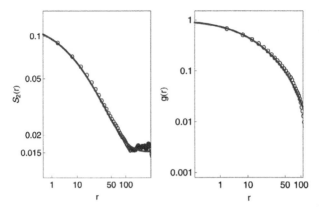

Fig. 4–6. (left) Two-point correlation function $S_2(r)$ as a function of distance r and (right) its normalized version $g(r) = [S_2(r) - \phi^2]/[\phi(1 - \phi)]$ computed for the soil image shown in Fig. 1.

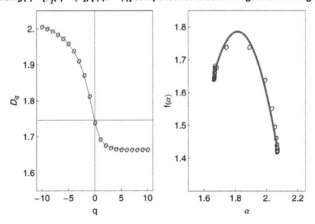

Fig. 4–7. Results of the multifractal analysis of the binary soil image shown in Fig. 1 (left) Fractal dimension D_q as a function of q and (right) singularity spectrum $f(\alpha)$ are shown.

these properties might also be valid for other network models, so in the SN network model for the optimal m

$$m_{\text{opt}} = \frac{D_f}{\alpha - 1}$$ [14]

and for the critical m

$$m_c = \frac{D + D_f}{\alpha - 1}$$ [15]

we find for the scaling exponent of the network degree distribution

$$\gamma_c = 2 + \frac{D_f}{D}$$ [16]

In the fractal model of the soil pore structure such as shown by Rieu and Sposito (1991) and Hunt and Ewing (2009) and Perfect et al. (2009), the probability density function for the pore radii follows a power-law with the exponent depending on the pore space fractal dimension: $W(r) \propto r^{-(1+D_f)}$ Using $s \propto r^D$ for pore sizes, we describe the pore size distribution by: $P(s) = W(r)dr/ds \propto s^{-(1+D_f/D)}$, so the scaling exponent of this distribution is

$$\alpha = 1 + \frac{D_f}{D} \tag{17}$$

This allows us to express the optimal

$$m_{opt} = D \tag{18}$$

and the critical m parameter of the SN network model of soil pore structure

$$m_c = D\left(1 + \frac{D}{D_f}\right) \tag{19}$$

as a function of the pore fractal and embedding dimension.

Large-scale organization of the network also has important consequences for network transport properties which is important in a soil biophysical sense for the movement of microbes and solutes. Scale-free organization of the network produces more efficient transport and communication between nodes (Lopez et al., 2005) than random network organization. Transport in scale-free networks (Lopez et al., 2005) was characterized with a distribution of the conductances between nodes $R(g) \propto g^{-\lambda}$, where the scaling exponent depends on the exponent of the scale-free network degree distribution $\lambda = 2\gamma - 1$. The conductance, g, was defined as the number of independent links between two nodes or as a function of node degrees of connected nodes (Lopez et al., 2005). Considering the optimal SN model (Eq. [14]) we include the conductance scaling exponent depending on the fractal dimension D_f of the soil porous architecture as $\lambda = 2\gamma_{opt} - 1 = 1 + 2D_f/D$.

Correlation Properties of Soil Porous Complex Networks

To correlate the fractal properties of the network structure with the porous structure obtained from the two-dimensional soil structure images, we consider the degree correlations k_{nn} or the assortativity or mixing of the network which is given by the Pearson coefficient, r, of the degrees of adjacent nodes (Newman, 2002). In the SN network model of soil pore structure we expect the mixing properties to depend on the parameter m. In geographical scale-free networks (with uniform distribution of nodes in Euclidean space) the crossover from disassortative ($r < 0$) to assortative ($r > 0$) was found to occur when the degree scaling exponent was γ

= 3 (Morita, 2006). This corresponds to the degree of scaling of average nodes in a growing EN network model. However, in the soil pore network model, we expect the crossover to occur at $\gamma = \gamma_c = 2 + D_f / D$ or at m_c given by Eq. [15]. In this case D_f = 1.74 corresponds with m_c = 4.3. The crossover point also depends on the value of the mean degree $\langle k \rangle$ (Johnson et al., 2010) and therefore serves as an additional constraint on the network model parameters. From the results displayed in Fig. 4–8 we can observe that the crossover from disassortative to assortative network structure increases from 4 to 4.5 as we increase the mean degree from $\langle k \rangle$ = 3 to $\langle k \rangle$ = 12. To have m_c = 4.3 in accordance with the fractal properties of the pore space, the mean degree should be $\langle k \rangle \approx 4$. The mean degree of the network corresponds to an average coordination number $\langle Z \rangle$ in pore networks obtained by morphological skeletonization (Liang et al., 2000) or to the effective coordination number $\langle <Z_{eff} \rangle$ in pore networks based on regular lattices.

Using Complex Networks to Generate Binary Images

Instead of using network theory to extract the soil pore structure from binary images as described above, we can create binary images with a growing network mechanism that statistically resemble images of the soil pore architecture derived from X-ray CT (Mooney and Korošak, 2009). This is important as the validity of binary images of soil structure derived from manual or automated thresholding algorithms is an area of concern between researchers in this area (see Baveye et al. (2010) for further discussion). Figure 4–9 displays the evolution of the binary image at two instances of simulation time. These were constructed with the heterogeneous preferential attachment mechanism (Eq. [7–8]) where all new nodes of

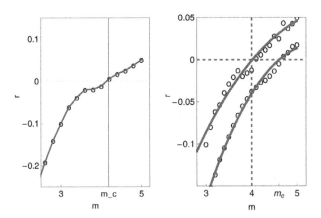

Fig. 4–8. Pearson's coefficient r applied to the links in the network as a measure of degree-degree correlations. (left) r as a function of parameter m in the static network model for $\langle k \rangle$ = 4 and (right) the influence of $\langle k \rangle$ on mixing properties of the network, the upper curve corresponds to $\langle k \rangle$ = 3, lower to $\langle k \rangle$ = 12.

Fig. 4–9. Binary image obtained with growing network algorithm simulating soil pore structure (black) at (left) early and (right) late stage of computation.

the network had equal intrinsic weights (setting $b = 0$ in Eq. [8]). In each step a new node j was introduced randomly into the two-dimensional Euclidean space and if connected to a randomly chosen existing node i (with the probability given by Eq. [7]), a pixel representing pore space was introduced into the binary image. The algorithm is stopped when the desired porosity of the simulated soil structure is reached which could be derived from soil structure images or by conventional laboratory measurements such as mercury porosimetry. The porosities of the presented images are $\phi = 0.1$ (left image in Fig. 4–9) and $\phi = 0.18$ (right image in Fig. 4–9). Figure 4–10 shows the pertinent normalized correlation functions using $g(r) \propto r^{-(D-D_f)} \exp(-r/\varsigma)$ and the singularity spectra $f(\alpha)$. The fractal dimensions at $q = 0$ are $D_f = 1.48$ for the image with $\phi = 0.1$, and $D_f = 1.62$ for the image with $\phi = 0.18$. This agrees well with the values obtained from fitting the correlation func-

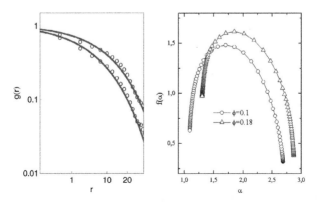

Fig. 4–10. Two point normalized correlation function $g(r)$ as a function of distance r of the simulated binary soil image from Fig. 9 (upper curve for porosity $\phi = 0.18$, lower curve for $\phi = 0.1$) and the corresponding singularity spectrum $f(\alpha)$ from multifractal analysis.

tion: $D - D_f = 0.6$ and $D - D_f = 0.5$. With increasing porosity also the cutoff length of the correlation function increased from $\zeta = 15$ to $\zeta = 18$.

Importantly, this example demonstrates that the binary images generated with the growing network mechanism simulating two-dimensional soil structure and two-dimensional images from real soil samples are statistically similar. In addition, both static and growing complex network representations of soil porous architecture points to similar scale-free network organization and reflect the structure of soil pore space such as multifractal spatial patterns which would cause multiscaling to appear in the network scaling exponent $\gamma(q) = \gamma[D_f(q)]$ also in the static network model. Multiscaling has also been connected to competition for links in growing networks where the connectivity of nodes depends on their fitness to acquire new links (Bianconi and Barabasi, 2001).

Conclusions and Outlook

The key difference between complex (SN or EN) pore network models and network models created on regular lattices, usually with a narrow distribution of coordination numbers, is that complex network models allow long-range links to exist between nodes and have nodes with a large number of links/hubs. The links in complex networks models do not necessarily represent true soil pore throats linking the pores, but rather describe the overall topological organization of the soil pore structure and provide clues to understand the relation of soil structure and function.

Given the relationship between the complex network structure and the spatial correlation properties of the soil pore structure as presented in examples in this chapter, we might wonder how can the suggested complex soil pore network organization help us to understand given soil functions such as the self-organization of soil–microbe complex (Young and Crawford, 2004). Here they suggested that depending on biological activity the local soil structure can regularly change between the more open (to enhance the rate of oxygen supply) and the more closed one (to protect the soil biological function). Therefore a key question to follow on from this is can we expect then the microbial activity and interactions with the soil structure to be reflected in the value of the network parameter m which on the other hand controls the navigability (Kleinberg, 2000) in a network with long-range connections? Does the self-organization of the soil–microbe complex result in the optimal network structure (Eq. [14]) in case of increased biological activity and causes the transition to a more compact network organization ($m > m_c$) with prevailing short-range links when life in soil needs to be protected?

References

Adler, P.M. 1985. Transport processes in fractals II. Int. J. Multiphase Flow 11:213–239. doi:10.1016/0301-9322(85)90047-3.

Adler, P.M., and H. Brenner. 1984. Transport processes in spatially periodic capillary networks. I. Geometrical description and linear flow hydrodynamics. Phys. Chem. Hydrodyn. 5:245–268.

Albert, R., and A.-L. Barabasi. 2002. Statistical mechanics of complex networks. Rev. Mod. Phys. 74:47–97. doi:10.1103/RevModPhys.74.47.

Barabasi, A.-L. 2009. Scale-free networks: A decade and beyond. Science 325:412–413. doi:10.1126/science.1173299.

Baveye, P.C., M. Laba, W. Otten, D. Grinev, L. Bouckaert, R.R. Goswami, Y. Hu, J. Liu, S.J. Mooney, R. Pajor, P. Dello Sterpaio, A. Tarquis, W. Wei, and M. Sezgin. 2010. Observer-dependent variability of the thresholding step in the quantitative analysis of soil images and X-ray microtomography data. Geoderma 157:51–63. doi:10.1016/j.geoderma.2010.03.015.

Berkowitz, B., and R.P. Ewing. 1998. Percolation theory and network model applications in soil physics. Surv. Geophys. 19:23–72. doi:10.1023/A:1006590500229.

Bianconi, G., and A.-L. Barabasi. 2001. Competition and multiscaling in evolving networks. Europhys. Lett. 54:436–442. doi:10.1209/epl/i2001-00260-6.

Bianconi, G., P. Pin, and M. Marsili. 2009. Assessing the relevance of node features for network structure. Proc. Natl. Acad. Sci. USA 106:11,433–11,438. doi:10.1073/pnas.0811511106.

Boccaletti, S., V. Latora, Y. Morenod, M. Chavez, and D.-U. Hwanga. 2006. Complex networks: Structure and dynamics. Phys. Rep. 424:175–308. doi:10.1016/j.physrep.2005.10.009.

Cardenas, J.P., A. Santiago, A.M. Tarquis, J.C. Losada, F. Borondo, and R.M. Benito. 2010. Soil porous system as heterogeneous complex network. Geoderma 160:13–21. doi:10.1016/j.geoderma.2010.04.024.

Dathe, A., A.M. Tarquis, and E. Perrier. 2006. Multifractal analysis of the pore- and solid-phases in binary two-dimensional images of natural porous structures. Geoderma 134:318–326. doi:10.1016/j.geoderma.2006.03.024.

Fatt, I. 1956. The network model of porous media, I. Capillary pressure characteristics. Petrol. Trans. AIME 207:144–159.

Freltoft, T., J.K. Kjems, and S.K. Sinha. 1986. Power-law correlations and finite size effects in silica particle aggregates studied by small-angle neutron scattering. Phys. Rev. B 33:269–275. doi:10.1103/PhysRevB.33.269.

Halsey, T.C., H.J. Mogens, L.P. Kadanoff, I. Procaccia, and B.I. Shraiman. 1986. Fractal measures and their singularities: The characterization of strange sets. Phys. Rev. A 33:1141–1151. doi:10.1103/PhysRevA.33.1141.

Horgan, G.W., and B.C. Ball. 1994. Simulating diffusion in a Boolean model of soil pores. Eur. J. Soil Sci. 45:483–491. doi:10.1111/j.1365-2389.1994.tb00534.x.

Hunt, A., and R. Ewing. 2009. Percolation theory for flow in porous media. Lect. Notes Phys. 771, Springer, Berlin.

Ioannidis, M.A., and I. Chatzis. 2000. On the geometry and topology of 3D stochastic porous media. J. Colloid Interface Sci. 229:323–334. doi:10.1006/jcis.2000.7055.

Johnson, S., J.J. Torres, J. Marro, and M.A. Munoz. 2010. Entropic origin of disassortativity in complex networks. Phys. Rev. Lett. 104:108702. doi:10.1103/PhysRevLett.104.108702.

Kleinberg, J.M. 2000. Navigation in a small world. Nature 406:845. doi:10.1038/35022643.

Lal, R. 2007. Soil science and the carbon civilization. Soil Sci. Soc. Am. J. 71:1425–1437. doi:10.2136/sssaj2007.0001.

Liang, Z., M.A. Ioannidis, and I. Chatzis. 2000. Geometric and topological analysis of three-dimensional porous media: Pore space partitioning based on morphological skeletonization. J. Colloid Interface Sci. 221:13–24. doi:10.1006/jcis.1999.6559.

Lopez, E., S.V. Buldyrev, S. Havlin, and H.E. Stanley. 2005. Anomalous transport in scale-free networks. Phys. Rev. Lett. 94:248701. doi:10.1103/PhysRevLett.94.248701.

Masuda, N., M. Hiroyoshi, and N. Konno. 2005. Geographical threshold graphs with small-world and scale-free properties. Phys. Rev. E 71:036108. doi:10.1103/PhysRevE.71.036108.

Mooney, S.J., and D. Korošak. 2009. Using complex networks to model two- and three-dimensional soil porous architecture. Soil Sci. Soc. Am. J. 73:1094–1100. doi:10.2136/sssaj2008.0222.

Morita, S. 2006. Crossovers in scale-free networks on geographical space. Phys. Rev. E 73:035104. doi:10.1103/PhysRevE.73.035104.

Newman, M.E.J. 2002. Assortative mixing in networks. Phys. Rev. Lett. 89:208701. doi:10.1103/PhysRevLett.89.208701.

Newman, M.E.J. 2003. The structure and function of complex networks. SIAM Rev. 45:167–256. doi:10.1137/S003614450342480.

Papadopoulos, A., N.R. Bird, S.J. Mooney, and A.P. Whitmore. 2009. Combining spatial resolutions in the multiscale analysis of soil pore size distributions. Vadose Zone J. 8:227–232. doi:10.2136/vzj2008.0042.

Perfect, E., Y. Pachepsky, and M.A. Martin. 2009. Fractal and multifractal models applied to porous media. Vadose Zone J. 8:174–176. doi:10.2136/vzj2008.0127.

Perrier, E., A.M. Tarquis, and A. Dathe. 2006. A program for fractal and multifractal analysis of two-dimensional binary images: Computer algorithms versus mathematical theory. Geoderma 134:284–294. doi:10.1016/j.geoderma.2006.03.023.

Petermann, T., and P. De Los Rios. 2006. Physical realizability of small-world networks. Phys. Rev. E 73:026114. doi:10.1103/PhysRevE.73.026114.

Peth, S. J. Nellesen, G. Fischer, and R. Horn. 2013. Dynamics of soil macropore networks in response to hydraulic and mechanical stresses investigated by X-ray microtomography. In: S. Logsdon, M. Berli, and R. Horn, editors, Quantifying and modeling soil structural dynamics. Advances in Agricultural Systems Modeling 3. SSSA, Madison, WI. p. 121–154. doi:10.2134/advagricsystmodel3.c6

Posadas, A.N.D., D. Giménez, R. Quiroz, and R. Protz. 2003. Multifractal characterization of soil pore systems. Soil Sci. Soc. Am. J. 67:1361–1369. doi:10.2136/sssaj2003.1361.

Prosperini, N., and D. Perugini. 2007. Application of a cellular automata model to the study of soil particle size distributions. Physica A 383:595–602. doi:10.1016/j.physa.2007.04.043.

Rieu, M., and G. Sposito. 1991. Fractal fragmentation, soil porosity, and soil water properties: I. Theory. Soil Sci. Soc. Am. J. 55:1231–1238. doi:10.2136/sssaj1991.03615995005500050006x.

Santiago, A., J.P. Cardenas, J.C. Losada, R.M. Benito, A.M. Tarquis, and F. Borondo. 2008. Multiscaling of porous soils as heterogeneous complex networks. Nonlinear Process. Geophys. 15:893–902. doi:10.5194/npg-15-893-2008.

Stanley, H.E., and P. Meakin. 1988. Multifractal phenomena in physics and chemistry. Nature 335:405–409. doi:10.1038/335405a0.

Tarquis, A.M., R.J. Heck, D. Andina, A. Alvarez, and J.M. Antón. 2009. Pore network complexity and thresholding of 3D soil images. Ecol. Complex. 6:230–239. doi:10.1016/j.ecocom.2009.05.010.

Vogel, H.-J. 1997. Morphological determination of pore connectivity as a function of pore size using serial sections. Eur. J. Soil Sci. 48:365–377. doi:10.1046/j.1365-2389.1997.00096.x.

Vogel, H.-J., and K. Roth. 2001. Quantitative morphology and network representation of soil pore structure. Adv. Water Resour. 24:233–242. doi:10.1016/S0309-1708(00)00055-5.

Watts, D.J., and S.H. Strogatz. 1998. Collective dynamics of "small-world" networks. Nature 393:440–442. doi:10.1038/30918.

Yeong, C.L.Y., and S. Torquato. 1998a. Reconstructing random media. Phys. Rev. E 57:495–506. doi:10.1103/PhysRevE.57.495.

Yeong, C.L.Y., and S. Torquato. 1998b. Reconstructing random media. II. Three-dimensional media from two-dimensional cuts. Phys. Rev. E 58:224–233. doi:10.1103/PhysRevE.58.224.

Yook, S.-H., H. Jeong, and A.-L. Barabási. 2002. Modeling the Internet's large-scale topology. Proc. Natl. Acad. Sci. USA 99:13382–13386. doi:10.1073/pnas.172501399.

Young, I.M., and J.W. Crawford. 2004. Interactions and self-organization in the soil–microbe complex. Science 304:1634–1637. doi:10.1126/science.1097394.

Chapter 5

A History of Understanding Crack Propagation and the Tensile Strength of Soil

Paul D. Hallett, Anthony R. Dexter, and Shuichiro Yoshida

Abstract

Despite the highly dynamic nature of cracks in soil and their significance for a wide range of processes including soil structure dynamics, few studies have attempted to quantify the physical processes involved in their genesis. In this chapter the development of fracture mechanics theories from materials science to describe crack propagation in soil will be discussed. It starts by considering some of the pioneering research on capillary bonding by Haines and Fisher that was developed for soil but has been extended to a wide range of particulate materials. By simplifying soil structure, this research showed how basic concepts can be developed and then extended to more natural systems. Soil mechanics research investigating the strength of unsaturated soil is then reviewed. In soil containing cracks, however, traditional unsaturated soil mechanics approaches have been found to provide poor predictions of soil strength. To incorporate the influence of cracks, various researchers have adopted linear elastic fracture mechanics. This approach assumes that all energy imparted to a soil system is recoverable on unloading, with only limited dissipation of plastic energy occurring. Later research applied the traditional J integral, which defines the elastic-plastic energy criterion for crack initiation. Simplified testing approaches and analysis have been developed recently that may allow for more widespread use of elasto-plastic fracture mechanics to describe crack initiation in soil. The need for future research on a wider range of soils, the coupling of fracture mechanics with soil hydrology, and advanced modeling approaches to predict crack depth are highlighted in this chapter. Such an understanding is essential for predicting the dynamic nature of soil structure.

Abbreviations: CTOA, crack tip opening angle; CTOD, crack tip opening displacement; LEFM, Linear Elastic Fracture Mechanics.

P.D. Hallett, The James Hutton Institute, Invergowrie, Dundee, DD2 5DA, United Kingdom (paul.hallett@hutton.ac.uk); A.R. Dexter, Institute of Soil Science and Plant Cultivation (IUNG), ul. Czartoryskich 8, 24-100, Pulawy, Poland (tdexter@iung.pulawy.pl); S. Yoshida, Dep. of Biological and Environmental Engineering, Graduate School of Agronomy and Life Science, Univ. of Tokyo, 1-1-1, Yayoi, Bunkyo, Tokyo, 113-8657, Japan (agyoshi@mail.ecc.u-tokyo.ac.jp).

doi:10.2134/advagricsystmodel3.c5

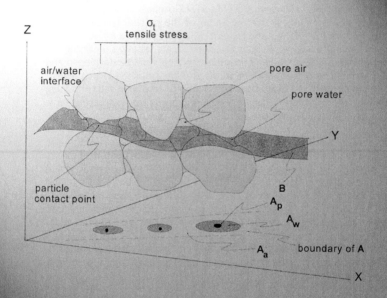

A s "the most complex biomaterial" on Earth (Young and Crawford, 2004), soil presents a fascinating challenge to researchers. Among the soil scientists who embrace complexity are the group who study soil structure. Here, we define soil structure as the complex arrangement of pores, particles, fluids, organic materials, and biota. In the early beginnings of the study of soil structure, two distinct camps of researchers formed: the micromorphologists interested in what we see (Kubiëna, 1938) and those interested in why we see it (Haines, 1925). Methodologies for the observation of soil structure have advanced considerably, with a significant recent leap provided by high resolution non-invasive imaging (Jaeger et al., 2009; Taina et al., 2008; Peth et al., 2010). Knowledge of the processes that lead to soil structure formation is good, with considerable effort placed in the study of properties of soil aggregates (Six et al., 2004). Our ability to predict how soil structure changes over time, however, is woefully inadequate, particularly when the magnitude of the potential impacts on food security, water transport/storage, and climate change are considered.

Many soil scientists bemoan that the major hindrance to progress is the combined complexity of the biological, physical, and chemical processes that underlie soil structure development and how it changes over time (Horn et al., 1994; Warkentin, 2008). Much of the work on soil structure development to date uses a laudable approach that attempts to preserve natural conditions, arguing that studies on repacked or model specimens create conditions too far removed from reality (Six et al., 2004). Soil aggregates and the pore spaces between them must be retained, even if they are sometimes volume elements plucked arbitrarily from larger units of soil. This approach varies considerably from historic soil physics and geotechnical engineering research, where the incorporation of soil structure was considered a step too far (Rahman et al., 2010; Zhang et al., 2008). Columns of sand or pure clay were the common approach used to develop theory in both disciplines.

It is now well accepted that soil structure has such a large impact on the transport properties and mechanical behavior of soil that it needs to be incorporated and understood (Chertkov and Ravina, 2004; Dexter, 2004; Roger-Estrade et al., 2009). However, much could be learned from the soil physicists and geotechnical engineers of yesteryear, who used model systems to build a fundamental understanding. Soil structure development requires understanding of bonding between soil particles (Raats, 1984), the impact of capillary stresses from water (Yoshida and Hallett, 2008), biological glues and lubricants (Chenu and Guérif, 1991), enmeshment of particles of roots (Mickovski et al., 2009) and fungi (Cavagnaro et al., 2006), shrinkage and swelling of clays (Chertkov, 2002), viscous flow of wet soil (Or, 1996), cracking of dry soil (Morris, 1992) and the wettability of surfaces (Ramirez-Flores et al., 2008). Incorporating understanding of all these processes simultaneously is virtually impossible, hence the qualitative nature of soil structure research. Rather than deconstructing natural soil structure, another approach used is to construct components of soil structure development to obtain a quantitative understanding (Haines, 1927; Or, 1996).

With this approach it is not only possible to more easily incorporate theory from soil physics and mechanics, but also from materials science. This is a massive discipline in comparison to soil science with almost 15 times as many publications in 2009 (Web of Science). Soil science has benefitted from adopting approaches from subdisciplines of materials science such as rheology, fracture mechanics, fiber reinforcement, and agglomeration. These have provided quantitative understanding that has been applied to model seedbed dynamics (de Brouwer et al., 2005), to bonding mechanisms between soil particles (Markgraf et al., 2006) and to the description of the reinforcement of soil by plant roots (Loades et al., 2010). This chapter focuses on how the study of cracking in soil has benefitted from bridging disciplines with materials science. There are clear challenges to adopting these approaches, including the simplification of soil structure complexity and the adaptation of testing approaches to work with soil specimens, but some of the examples provided will demonstrate major leaps in understanding.

Early Pioneers—Haines and Fisher

Water is one of the major drivers of soil structure development as it impacts soil strength, shrinkage, swelling and cracking. In the 1920s Keen (1924) demonstrated that water contributes to the tensile strength of soil and this was followed up by Haines (1925) in pioneering research to derive formulae to calculate the extent of this force. Although his interest was soil, he considered capillary forces acting between two spheres resulting from diminished pressure within a meniscus. This simplification of soil into a model structure has allowed the fundamental processes involved to be disentangled. Haines (1925) research was complemented by research by Fisher (1926) who included the influence of surface tension to present the first quantitative theory of capillary attraction between spherical particles. Not only was this a significant advance in soil science, but also in the study of any particulate system including chemical powders. The derivation of the theory is shown below to demonstrate its progression from an ideal system to an approach that can be used on intact soil.

With the use of a model of an ideal soil consisting of a toroidal bridge of water between two incompressible, touching spheres (Fig. 5–1), Fisher (1926) calculated that the tensile force, F_{ST}, exerted by the gas–liquid interface is

$$F_{ST} = 2\pi r_2 \gamma_w \qquad [1]$$

where r_2 is the radius of the neck of the bridge, $2\pi r_2$ is the perimeter at the neck and γ_w is the surface tension of the air–water interface.

Fisher (1926) also determined that the tensile force, F_c, due to the diminished pressure, p_w, within the liquid is

$$F_c = \pi r_2^2 \gamma_w \left(\frac{1}{r_1} - \frac{1}{r_2} \right) \qquad [2]$$

where r_1 is the radius of curvature of the water bridge (Fig. 5–1). The forces due to capillarity and surface tension can be combined to give the total tensile force due to pore water as

$$F_T = F_{ST} + F_c = 2\pi r_2 \gamma_w + \pi r_2^2 \gamma_w \left(\frac{1}{r_1} - \frac{1}{r_2} \right) \qquad [3]$$

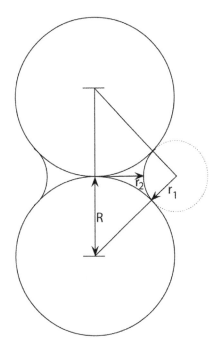

Fig. 5–1. Idealized model of a bridge of capillary water between two incompressible, touching spheres where R is the radius of the sphere and r the radius of the capillary bridge.

Haines' (1927) further work on how capillary bonds influence the tensile strength of adjoining spheres sparked a highly significant discovery in soil physics. The theory presented so far has only considered independent bridges of pore water, so Haines examined the effect of coalescence of pore water on pressure deficiency. He impregnated model systems of closest packed spheres with wax to visualize the distribution and angle of curvature of the water films at different water contents. Haines concluded that coalescence causes the water films to readjust causing an abrupt rise in pressure deficiency. Before coalescence, the pressure deficiency, p_w could be calculated as

$$p_w = \frac{(6.4 - 2.3)\gamma_w}{R} \qquad [4]$$

where R is the radius of the sphere. After the water films are rearranged at coalescence, the pressure deficiency approaches

$$p_w = \frac{(6.4 + 6.4)\gamma_w}{R} \qquad [5]$$

The values of 6.4 and 2.3 were obtained from the estimated volume of air and water in the arrangement of spheres. These were measured directly by embedding the model sample in wax to measure the air volume. Equation [5] was interpreted by Haines (1927) to represent the "air-entry" pressure, which is the minimum pressure deficiency that an air–water interface can develop in soil. The abrupt change in pressure deficiency that Haines (1927) modeled in Eq. [4] and [5] is not seen for real soils. Most soils shrink to some extent on drying, which Haines (1927) hypothesized buffered any effects of coalescence.

Fisher (1928) found faults with this analysis and in particular noted that Eq. [5] predicts far too high a static stress. The approach he adopted compared the pressure excess over the dry portion of the ideal soil with the pressure deficiency over the wet portion. The pressure in the wet portion is simply p_w, while at the interface of the dry portion it is $2\gamma_w/R$, resulting in a total pressure of $p_w + 2\gamma_w/R$. With this knowledge it was possible to describe the strength of water bonds from the area of wet and dry surfaces alone.

From these studies on glass balls, pivotal leaps in understanding of soil physics were made possible. The concept of an air-entry pressure was derived and the fundamental physical processes underlying interparticle bonding by pore water were described in elegantly simple equations based on fundamental physical properties. Decades later, Terzaghi (1943) revisited the contribution of pore water to the bonding of particles and strength of soil. He developed the concept of effective stress, σ', which for saturated soils is

$$\sigma' = \sigma + (\mu_a - \mu_w) \tag{6}$$

where σ is the externally applied stress, μ_a is the air pressure and μ_w is the pore water pressure. On the basis of Haines' (1927) and Fisher's (1928) earlier research, Eq. [6] can be adapted for unsaturated soils by accounting for the degree of pore saturation, S as

$$\sigma' = \sigma + S(\mu_a - \mu_w) \tag{7}$$

This simple relationship is still used to obtain approximate solutions for the stress contributed by pore water in soil (for example Mullins and Panayiotopoulos, 1984). In experiments on soil, however, the relationship between an applied stress to soil and effective stress is nonlinear (Aitchison, 1957). Bishop (1959) proposed that S be replaced by a parameter, χ, which could account for pore geometry along the failure surface in soil. χ in this instance was evaluated by shear tests at different pore water pressures. An interpretation of χ presented by Aitchison (1961) considers it to be a weighted average of the size of individual soil elements that form the failure surface. χ can be expressed in the basic form of Eq. [7] as

$$\sigma' = \sigma + \chi(\mu_a - \mu_w) \tag{8}$$

This equation to describe the effective stress of soil underlies much of modern day unsaturated soil mechanics. The parameter χ is essentially a fitting parameter to account for pore water distribution at the failure surface, so often it fails to describe adequately the relationship between water potential and soil strength. Blight (1967) noted that χ is not independent of the mode of loading and is changed by volumetric strain. Although capillary theory dictates that the pore water pressure becomes less negative with particle separation, volumetric strain may in fact redistribute the pore water and thus increase χ over a limited range so that the effective stress is increased.

It has been suggested that for structured soils, χ as it is traditionally defined should be replaced by an inter-aggregate effective stress parameter, χ_a (Snyder and Miller, 1989). Nearing (1995) successfully applied this concept in evaluating the compressive strength of soil aggregate beds. Extending the basic effective stress equation (Eq. [8]) to account for soil structure results in

$$\sigma' = (\sigma - \mu_a) + \chi_a(\mu_a - \mu_w) \tag{9}$$

By using the data and original theory of capillary cohesion of Haines (1927), Snyder and Miller (1985) developed a theoretical relationship for χ. They treated soil as a composite of ideal spherical particles, as defined by Keen (1924) of different radius and derived the following,

$$\chi = S + \frac{0.3}{\mu_a - \mu_w} \int_S^1 (\mu_a - \mu_w)_i dS_i \qquad\qquad [10]$$

where S_i and $(\mu_a - \mu_w)_i$ are the degree of pore saturation and pore water pressure deficiency, respectively, at any point i between S and complete saturation. This equation provided an improvement in the description of the contribution of pore water to the strength of soil. When extended further by Snyder and Miller (1985) to predict cracking in structured soils, a reasonably accurate agreement between experimental and theoretical values was found. This will be described in greater detail in the crack mechanics section.

The basic concept of χ has been applied in many papers that examine the mechanical behavior of unsaturated soils (for example Farrell et al., 1967; Williams and Shaykewich, 1970; Mullins and Panayiotopoulos, 1984). By building up complexity using model systems of soil, various alternative solutions to describe effective stress have been developed (Groenevelt and Kay, 1981; Snyder, 1987; Pande and Pietruszczak, 1990). Chemical engineers have also studied capillary cohesion to understand the behavior of particle agglomerates (Lian et al., 1993; Simons et al., 1994).

Although it is recognized that capillary cohesion contributed by pore water can be the most significant strength-controlling property in soil, the first principles set out by Haines (1927) need to be remembered. By starting with an elegantly simplistic geometry to mimic soil, he provided an understanding of capillary cohesion that has applications far beyond the study of soil. A great challenge remains in extending this theory to naturally structured soils, although Snyder and Miller (1985) and other researchers have demonstrated the possibilities.

Understanding Soil Cracking

Soil structure is a product of dispersion, agglomeration, and cracking. From pictures of crack shapes in drying soil or data on the stability of soil aggregates, it has been demonstrated that a vast range of biological, chemical and physical processes underlie both processes. This is interesting and important in developing sustainable solutions to the management of soil, but we believe that more effort should be diverted back into understanding the basic principles. Models to describe soil structure formation or dynamics are scarce, whereas thousands of papers on soil aggregates would benefit from this understanding.

Cracking in soil is governed by the resistance contributed by the forces of the various bonds that hold the soil particles together along the fracture surface, the area over which these forces act and external forces applied to the soil (Gill and

Vanden Berg, 1968). This concept is illustrated in Fig. 5–2, which shows an unsaturated soil element transversed by an imaginary cross-sectional surface of area, A. The wavy plane, which ultimately defines a crack surface, intersects an area of water, A_w, air, A_a, and interparticle contacts, A_p, which for illustrative and computational purposes have been projected onto the x–y plane. Snyder and Miller (1985) evaluated the force balance in the z direction as

$$\sigma_i A = F_p + A_w \mu_w + A_a \mu_a - \gamma \int_0^P \mathbf{i} \times \mathbf{j}\, ds \qquad [11]$$

where F_p is the z component of the sum of interparticle forces acting across the surface B. The z component of the surface tension force, γ, is evaluated by the integral term on the right of Eq. [11], where $\mathbf{i}$ is the unit vector in the z direction and $\mathbf{j}$ is the unit vector tangential to the air–water interface and perpendicular to a section ds along which the surface B intersects the air–water interface. P is the sum of the circumferences of the air–water interfaces that are intersected by the surface B. The three component forces due to pore water were first presented in a form similar to Eq. [11] by Sparks (1963).

The stress balance is obtained by dividing by A to obtain,

$$\sigma_t = \frac{F_p}{A} + \frac{A_w}{A}\mu_w + \frac{A_a}{A}\mu_a - \frac{\gamma}{A}\int_0^P \mathbf{i} \times \mathbf{j}\, ds \qquad [12]$$

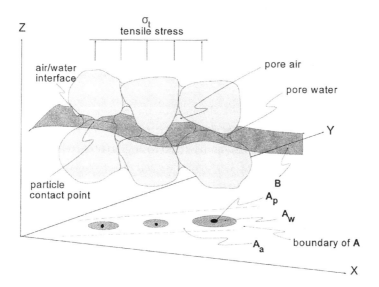

Fig. 5–2. The various bonds that are broken during tensile failure of an unsaturated soil element having the projected boundary, A, transversed by an imaginary surface, B, where the subscripts a, w, and p refer to air, water and particle, respectively (after Snyder and Miller, 1985).

Through further derivation, Eq. [12] can include χ, inevitably allowing it to be reduced to the simple effective stress theory described in Eq. [8].

One application of this approach to describe cracking in soil by tensile failure was conducted by Mullins and Panayiotopoulos (1984) who tested the mechanical strength of unsaturated mixtures of kaolinite and sand equilibrated to selected initial negative pore water pressures. To calculate how pore water pressure influences the tensile failure stress, σ_f, they predicted that

$$\sigma_f = c + S\mu_w \qquad\qquad [13]$$

where c is the cohesion contributed by soil bonds, as determined using Mohr–Coulomb theory. They found an "encouraging" relationship between the predicted and measured tensile failure stress from their experiments on unsaturated mixtures of sand and kaolinite. These tests were not confounded by the complexity of soil structure as the samples were all reconstituted. Young and Mullins (1991) applied Eq. [13] to undisturbed cores of soil. The parameter c was omitted, thus relying on the water potential as the strength-controlling property of soil. At an initial water potential of –1000 kPa, they found that the theoretically predicted values were considerably greater than the measured values. It was concluded that at very negative water potentials, failure would be dominated by brittle fracture which makes Eq. [13] unsuitable. Further, they recognized that the area of the water films along the fracture surface could be much smaller than what would be calculated from the degree of saturation of the bulk soil. These limitations were recognized by Mullins et al. (1992) who suggested that the effect of u_w on σ_f be evaluated using the relationship

$$\sigma_f = i + s(S\mu_w) \qquad\qquad [14]$$

where the coefficients i and s are obtained by linear regression and correspond to c and the reciprocal of the pore shape factor, respectively. The latter parameter accounts for the stress intensification defined by Inglis (1913), which is discussed in the next section. Equation [14] provided a good fit between experimental data collected for naturally structured soil and theory. This represented a significant leap in predicting cracking and the strength of naturally structured soils. Without the initial research by Mullins and Panayiotopoulos (1984) on idealized mixtures of sand and kaolinite, however, Eq. [14] might never have been derived.

Although the fitting parameters i and s in Eq. [14] have a sound physical basis, it is not possible to determine if they in fact represent cohesion and stress intensification accurately. If advancements are to be in the understanding rather than prediction of soil fracture, adaptations to classical soil mechanics such as Eq.

[13] need to embrace advances from other disciplines. In particular, theoretical advances in fracture mechanics may enable a much greater understanding of the dynamics of soil structure.

Fracture Mechanics

Soil science has been slow to adopt approaches from fracture mechanics. This discipline originated following a range of disasters involving man-made structures, particularly hull ruptures in trans-Atlantic shipping vessels. Engineers were puzzled for years why existing theories to describe the strength of metals provided poor predictions, but work by Inglis (1913) discovered that the concentration of stresses at the tip of cracks reduces strength considerably (Fig. 5–3). With the assistance of his mathematical work, Griffith (1921) derived the theoretical conditions required for a crack to propagate in a brittle material. He determined that in an "ideal solid" (i.e., isotropic and perfectly elastic), the energy associated with crack formation, U, is the sum of the mechanical energy transmitted to the material, U_m, and the free energy expended in creating new crack surfaces, U_s. Hence,

$$U = U_m + U_s \tag{15}$$

Up until crack failure, U_m is stored as potential energy in the system, U_E. By calculating the stress and strain fields near to the crack as determined by Inglis (1913) and integrating over dimensions much greater than crack length, c, Griffith found for a unit width along the crack front

$$U_E = \frac{\pi c^2 \sigma_A^2}{E} \tag{16}$$

where E is the Youngs modulus and σ_A is the applied stress. The surface energy for unit width of crack front is

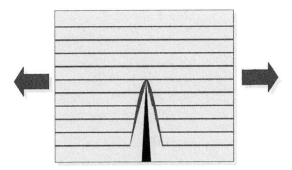

Fig. 5–3. Conceptual diagram of how a stress is intensified in a material containing a crack. The lines represent the distribution of stress applied in tension (shown by arrows). As stresses do not transfer across a crack, they become concentrated at the tip.

$U_s = 4c\gamma_s$ [17]

where γ_s is the free surface energy per unit area and the value 4 arises because the crack is of length $2c$ and 2 surfaces are separated. The total system energy is the sum of Eq. [16] and [17],

$$U(c) = \frac{-\pi c^2 \sigma_A^2}{E} + 4c\gamma_s$$ [18]

As a crack extends in an elastic material, the mechanical energy required decreases ($dU_m/dc < 0$), while the surface energy increases ($dU_s/dc > 0$). Catastrophic failure conditions result when, $dU/dc = 0$ [hence $U(c) = 0$], which is known as the Griffith criterion for fracture (Fig. 5–4). At failure, $\sigma_f = \sigma_A$, so combining Eq. [15] to [18] results in,

$$\sigma_f = \left(\frac{2E\gamma_s}{\pi c} \right)^{1/2}$$ [19]

The crack energetics can be conveniently related to the surface energy by defining a quantity called the mechanical energy release rate, G,

$$G = \frac{dU_m}{dC}$$ [20]

where C is the crack interfacial area (Griffith, 1924). Irwin (1958) found that G could be used to define the intensity of stress near to the edge of a crack. He derived the concept of a stress intensity factor, K, evaluated as

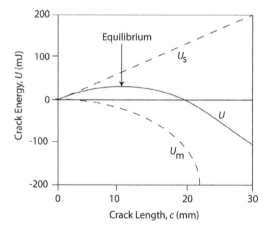

Fig. 5–4. Energy balance of a crack in uniform tension, plane stress, based on Griffith theory. Data for glass from Griffith (1921): surface energy, γ = 1.75 J m^{-2}; Young's Modulus, E = 62 GPa; applied tensile stress, σ_A = 2.63 MPa (chosen to give equilibrium at c = 10 mm) (redrawn from Lawn, 1993). U is the total energy, U_s is the surface energy, and U_m is the mechanical energy.

$$K = \left(\frac{GE}{\pi}\right)^{1/2}$$
[21]

This revolutionized fracture mechanics since K is very easy to evaluate using simple tests. The critical condition for fracture of a cracked specimen can be described by

$K = Y\sigma c^{1/2}$
[22]

where Y is a dimensionless term which accounts for specimen geometry. The ability of an elastic material to tolerate a crack is defined by its critical stress intensity factor or fracture toughness, K_c, with failure occurring if $K \geq K_c$. The equations above are referred to as Linear Elastic Fracture Mechanics (LEFM) because they assume that all imparted energy is either recoverable or used to create new fracture surfaces. The work described in the following section generally assumes LEFM behavior.

Applying Linear Elastic Fracture Mechanics to Understanding Cracking in Soils

Fippin (1910) recognized that cracks were significant in the fracture of soil. He discussed soil fracture using cohesion theory, similar to the approaches used originally to determine the strength of trans-Atlantic shipping vessels. In a comprehensive report on soil structure, Russell (1938) emphasized the importance of cracks and was among the first to realize the hierarchical nature of aggregates in soil. Tensile failure was considered by Richards (1953) and Kirkham et al. (1959) who suggested that the modulus of rupture be measured since it is a "true" physical property. Ingles (1963) and Ingles and Frydman (1963) recognized crack propagation as the primary failure mechanism in dry soil but commented that understanding was confounded by a lack of suitable testing procedures that would yield reproducible results. The test suggested by Ingles (1963) was a drop shatter test taken from the coal industry. In this test, the sample is dropped from a prescribed height for as many times as is required for fracture to occur. The cumulative drop distance is then used to evaluate the kinetic energy required for fracture. This test not only suggests that soil fails by crack propagation, but that cracks grow and arrest even in dry soil and that some energy is stored. The tests suggested by Ingles and Frydman (1963) were conventional procedures used in materials testing adapted through the use of remolded soil samples formed to specific shapes and dimensions.

Results obtained by Ingles and Lafeber (1966) indicated that crack development in soil was most probably a factor of brittle fracture. The following year, they published the first comprehensive review in which it was suggested that soil fracture be examined using Griffith theory rather than a classical soil mechanics approach (Ingles and Lafeber, 1967). In the same year, Farrell et al. (1967) presented a model that used Griffith theory to evaluate the effects of cracks on the fragmentation behavior of a model soil. This work was never published which is unfortunate since it could have served as a useful foundation for future research.

Unlike even more recent work on soil fracture (Mullins and Panayiotopoulos, 1984;Hallett et al., 1995;Hallett and Newson, 2001, 2005), both Ingles and Lafeber (1967) and Farrell et al. (1967) recognized soil as an extremely complex structural material. Both papers discussed a model presented by McClintock and Walsh (1962) for rock fracture which assumed that while some cracks would elongate under stress, other cracks would close in the compression zone causing friction. McClintock and Walsh (1962) modified the Griffith equation (Eq. [19]) for triaxial loading conditions to

$$f = \left(\sigma_1 + \sigma_3 - 2\sigma_c\right) - \left(\sigma_1 - \sigma_3\right)\sqrt{1 + f^2} = -4\sigma_t\sqrt{1 + \sigma_c/\sigma_t} \qquad [23]$$

to include the coefficient of friction, f, the principal stresses σ_1, σ_3, and the normal stress required to close, σ_c, and open σ_t the surfaces of Griffith cracks Fig. 5–5 shows the stress–strain relationship for a stabilized Montmorillonite clay tested in unconfined compression by Ingles and Lafeber (1967). They interpreted these results using a series of hypotheses based on previous findings in soil and rock mechanics. First, there is an evident increase in Young's Modulus at small strains, which they attributed to the closing of macrocracks that are oriented parallel to the applied load. With increasing strain, another deviation in the curve occurs

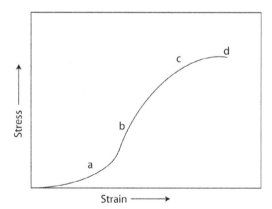

Fig. 5–5. Step-strain phenomenon in stabilized clays: closing of macrocracks (a), propagation of microcracks (b), weakening of material by linking of microcracks (c), and eventually ultimate failure (d) (lines sketched from the general relationship observed by Ingles and Lafeber (1967) for Montmorillonite loaded in unconfined compression).

from the propagation of microcracks. These microcracks finally link into incipient failure surfaces causing material weakening and a departure from linearity.

Farrell et al. (1967) commented on the effect of confining pressure on the stress at the crack tip. They referred to a model by Niwa and Kobayashi (1967) which assumes that the maximum tensile stress at a crack tip is reduced by a factor of $1/(1 + m)$ by a confining pressure. This criterion, given that m is a material constant, could be related to their experimental data for mortar as

$$\left(\sigma_1 - \sigma_2\right)^2 - 8\sigma_t\left(1 + m\right)\left(\sigma_1 + \sigma_2\right) + 64\sigma_t^2 m = 0 \qquad [24]$$

Finally, Farrell et al. (1967) considered the effect of crack interactions on the failure stress. Essentially, cracks will interact when the stress fields at their tips overlap. There are complex equations to describe the stress field, but Yokobori et al. (1965) provided a simplified empirical rule that two cracks separated by less than 10% will behave as a single crack.

Ingles and Lafeber (1967) and Farrell et al. (1967) provided ground-breaking hypotheses of soil fracture. It was not until two decades later, however, that attempts were made to model the observed fracture behavior of soil and to analyze the results using Griffith theory.

A comprehensive model of unsaturated soil fracture that includes the basic principles of fracture mechanics was presented by Snyder and Miller (1985). This model is based on the concentration of stress at the crack tip and effective stress contributed by pore water as defined by Aitchison (1961). The stress concentration was taken as being dependent on crack shape rather than size. This is the same approach originally presented by Inglis (1913) in which the stress concentration for an elliptical crack was calculated as

$$\frac{\sigma_m}{\sigma} = 1 + 2\frac{c}{b} \qquad [25]$$

where c is the half length and b the half width of the ellipse, and σ_m is the maximum stress at the edge of the crack. σ_m/σ was selected over Griffith's strain energy approach since it requires evaluation of only σ_m rather than E, γ_s, and c. σ_m/σ for infinitely sharp cracks is mathematically equivalent to K defined in Eq. [21].

Snyder and Miller (1985) defined a parameter $f(\theta^*)$, by

$$\frac{\left(\sigma - \mu_a\right)_m}{\left(\sigma - \mu_a\right)} = f\left(\theta^*\right) \qquad [26]$$

to account for the intensification of the applied load due to the shape of the cracks. They hypothesized that the geometry of the crack trip would be controlled by the curvature of the meniscus which is a function of the degree of pore saturation.

The condition for failure defined by Snyder and Miller (1985) for an unsaturated soil was that the effective stress at the crack tips be equal to zero. Referring to Eq. [8], the value of $(\sigma - \mu_a)_m$ at rupture can be calculated by

$$\chi = \frac{(\sigma - \mu_a)_m}{(\mu_a - \mu_w)} \qquad\qquad [27]$$

By substituting Eq. [27] into Eq. [26] the failure criterion could be related to the applied stress $(\sigma - u_a)$ by

$$\chi / f(\theta^*) = \frac{(\sigma - \mu_a)}{(\mu_a - \mu_w)} \qquad\qquad [28]$$

Snyder and Miller (1985) then deduced that since both χ and $f(\theta^*)$ are functions of the degree of pore saturation, this equation could be generalized to

$$F(\theta^*) = \frac{(\sigma - \mu_a)}{(\mu_a - \mu_w)} \qquad\qquad [29]$$

where

$$F(\theta^*) = \chi / f(\theta^*) \qquad\qquad [30]$$

Snyder and Miller (1985) tested this equation using the results of Vomocil et al. (1961), Farrell et al. (1967), and Snyder (1980). When the effects of stress intensification by cracks were ignored, they found a considerable overestimation of predictions. If they assumed the stress concentration for a circular crack [$f(\theta^*)$ = 2], calculated values of failure stress approached the upper limit of experimental data they had for remolded silts and find sands. When applied to structured aggregated soil, the model greatly overpredicted strength. Snyder and Miller (1985) attributed this to the complexity of stress concentration and χ on the fracture surface of naturally structured soil. Nevertheless, their model represented a major advance in understanding.

One approach to deal with stress intensity in naturally structured soils was developed by Guérif (1988). He suggested that by testing aggregates of different sizes, it was possible to discard the effects of inter-aggregate pores. He hypothesized that aggregates 2 to 3 mm in diameter had only textural intra-aggregate pore space. By comparing the strength of aggregates of this size with larger aggregate size, it was possible to evaluate the influence of structural pores, something Guérif (1990) later concluded was a major strength-controlling property of soil.

Fracture mechanics was also used to describe crack propagation in desiccating soil by Nieber (1981) and Raats (1984). Nieber (1981) presented a series of images depicting crack propagation over time in a desiccating soil from which it was recognized that cracks grow from preexisting microcracks. This characteristic cracking pattern was described by Raats (1984) using a hypothesis presented by Lachenbruch (1961, 1962) that cracks are stress free along their walls. Raats (1984) commented that further progress in understanding these mechanisms could be expected from the combination of the usually independent disciplines of soil physics, geotechnical engineering, rock mechanics and fracture mechanics.

Among the first studies in soil science to apply fracture mechanics to soil was that conducted by Lima and Grismer (1994). They observed great changes in the cracking of soils affected by salinization and were interested in understanding the underlying mechanisms. Tests on saline versus unaffected soils found a doubling of G_c. Murdoch (1993a) made a direct attempt to use fracture mechanics to describe soil cracking using Eq. [21]. This study considered the hydraulic fracturing of soil, which is a common procedure used in rock mechanics to extend preexisting cracks (Murdoch, 1993b). It was found that K_c when evaluated for soil could be used to predict the conditions required for crack propagation. K_c was found to decrease with increasing water content, which was also found in a different study on soil by Nichols and Grismer (1997). For different initial crack lengths, however, values of K_c were different, suggesting that wet soil may be too nonlinear for LEFM to be appropriate.

This impact of ductility on the applicability of LEFM to soils was investigated further by Harison et al. (1994). Like Murdoch (1993a), they found that soil water content had the greatest impact on K_c. The driest soils had values of K_c that were 5 to 10 times greater than the wettest soils. Between different types of soil, the greatest differences could be attributed to the plasticity index. Ductility was assessed by Harison et al. (1994) from the ratio of fracture toughness to tensile strength, K_c/σ_t. Water was found to have a similar effect on soil ductility as temperature does for metals. They adopted a common procedure used in metals classification in which ductility was divided into three regions. In soils these are defined by the water content, w at the shrinkage limit, w_{sl} and plastic limit, w_{pl} of soil: (i) Region I, brittle ($w \leq w_{sl}$); (ii) Region II, transition water content range ($w_{sl} < w < w_{pl}$); and (iii) Region III, ductile ($w \geq w_{pl}$).

From the research described so far, advances have been made that suggest fracture mechanics provides a very valuable tool to describe and predict crack growth in soils. The great challenges are the complexity of the inherent pore structure of soil, the direct impact of capillary cohesion from water and the impact of ductility in wet soils. More recently attempts have been made to address these challenges, in addition to direct attempts to use fracture mechanics to describe soil cracking depths and patterns that occur in the field.

Nonlinear Fracture Mechanics of Soil

Even for dry soils, Hallett et al. (1995) argued that conventional LEFM may not be appropriate because of the amount of plastic energy lost during fracture. They adopted an approach from powder mechanics (Mullier et al., 1987) in which a crack length increment, Δc was added to the length of the initial crack, so the critical stress intensity factor according to Eq. [22] became

$$K_c = Y\sigma_t (c + \Delta c)^{1/2} \tag{31}$$

to account for the size of "process zone" in advance of the crack tip. Specimens of remolded soils with different inserted crack lengths were tested to evaluate K_c. They found that K_c was affected by initial crack length if Eq. [22] was used, but that Eq. [31] rectified the problem. The use of deep-notched specimens to minimize the influence of Δc over c was proposed as a simple method to evaluate K_c more accurately. Although Δc was argued to account for interparticle friction and deformation in advance of a crack, its physical meaning is tenuous and the results they obtained were about 1 order of magnitude greater than for powder agglomerates tested by other researchers (Ennis and Sunshine, 1993; Adams et al., 1989; Mullier et al., 1987).

A more robust approach may be provided by nonlinear fracture mechanics. The total energy for fracture, U incorporates both elastic energy, U_{el} and plastic energy, U_{pl},

$$dU = dU_{el} + dU_{pl} \tag{32}$$

to account for processes such as large scale irrecoverable plastic deformation of materials during fracture. Fracture toughness, K_c is replaced by the J integral to account for energy expended both in creating new surfaces, plastic processes and crack elongation:

$$J = \frac{\eta U}{Bb} \tag{33}$$

where U is taken from the area under the loading curve, B is the thickness of the specimen, η is a dimensionless constant, and b is length of material in advance of the crack. For pure bending in specimens with deep cracks, $\eta = 2$. J evaluated by Eq. [33] is the rate of energy release in plastic materials up to the critical condition when energy release causes unloading.

At this condition the energy absorbed by the specimen is calculated with steps of incremental loading, i as:

$$J_R = \sum_{i=1}^{N} \frac{2(P_{i-1} + P_i)\Delta q_i}{B(b_{i-1} + b_i)} \tag{34}$$

where P is the force, q is the load-point displacement and subscript i denotes the value of the ith incremental step. The subscript R indicates the resistance to crack growth after initiation by the material. Figure 5–6 provides an example for soil of the change in J with crack extension, which is commonly referred to as a J-R curve in materials science. The curve is characterized by an initial steep rise followed by a linear slope and final upward deviation. The first steep rise represents the blunting stage of the predefined notch. The force applied to the specimen was still rising in this stage (Fig. 5–6). The second linear rise corresponds to stable crack growth, where the force rapidly decreased. The third part represents the crack almost reaching the loading point, which is the limit of sample compliance for testing.

Chandler (1984) made the first attempt at applying nonlinear fracture mechanics to soil. Information on the imparted energy versus crack growth is required, which he measured on bend specimens by applying a voltage and estimating crack length from the increase in effective electrical resistance. Surprisingly, Chandler (1984) found that his application of nonlinear fracture mechanics to soil was insensitive to soil composition, which he attributed to the vast amount of energy expended in plastic deformation. When large-scale ductile crack growth occurs,

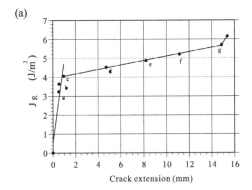

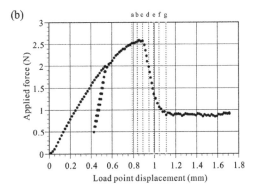

Fig. 5–6. The relationship between (a) the energy release rate, J_R and crack extension and (b) the corresponding points for applied force and load point displacement. Reprinted from Yoshida and Hallett (2008). Reproduced by permission of American Geophysical Union.

an alternative approach is to measure the crack tip opening displacement (CTOD) or crack tip opening angle (CTOA). This approach is widely adopted in materials science since it is easy to apply and provides useful parameters for predicting the strength of structures (Newman et al., 2003). Purists debate the physical meaning of the derived parameters, but work by Turner and Kolednik (1997) demonstrated a simple test method and the relevance of the derived parameters to quantifying crack propagation energetics.

CTOA was first applied to investigate crack propagation in soil by Hallett and Newson (2001) who adapted a simple deep-notch bend test developed by Turner and Kolednik (1997). Wet soils are incredibly weak so conventional bend tests are affected by self weight, which was minimized by balancing the specimen on the loading rollers in the bend test apparatus (see Fig. 5–7).

CTOA or $\alpha_{g,pl}$ describes the relationship between crack opening and extension. It could therefore be useful in soil science for predicting crack growth in shrinking soils. Data on load point displacement, q, span length of the loading rollers, S, crack length, c, and length of soil being flexed in advance of the crack, b is obtained from the bend test to evaluate $\alpha_{g,pl}/r^*_{pl}$ as

$$\frac{dq_{pl}}{da} = \frac{S\alpha_{g,pl}}{4r^*_{pl}b} \tag{35}$$

where r^*_{pl} is the distance ahead of the crack where the center of rotation occurs. This relationship assumes bending is fully plastic, hence the subscript pl. To obtain $\alpha_{g,pl}$, r^*_{pl} is calculated from crack mouth opening (measured by image analysis), V_{pl} and q_{pl} as

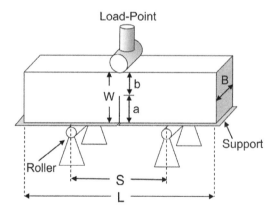

Load-Point

W

b

a

B

Support

Roller

S

L

Fig. 5–7. Deep-notch three-point bend test showing the specimen length, *L*, width, *W*, thickness, *B*, crack length, *a*, and ligament length, *b*. The specimen is supported on rollers spaced at distance *S* (*S* = *L*/2) and bent by downward displacement at the load point where force is measured. Reprinted from Yoshida and Hallett (2008). Reproduced by permission of American Geophysical Union.

$$r_{pl}^* = \left[(S/4)\left(dV_{pl}/dq_{pl}\right) - a_0 \right]/b_0 \qquad [36]$$

where a_0 and b_0 are the initial values of the crack and ligament (length of material ahead of the crack) lengths. CTOA was found to be very sensitive to both salinity and sand content in remolded (Hallett and Newson, 2001) and consolidated (Hallett and Newson, 2005) test specimens formed from kaolinite and fine sand.

Yoshida and Hallett (2008) used the CTOA approach to investigate the impact of hydraulic stress on the fracture of montmorillonite-sand mixes and a paddy soil. The aim of this work was to determine the cracking behavior of paddy soils in the field, as cracks are vital to drainage before harvesting. Through the use of image analysis of the growth of individual cracks, the local, α_t rather than global $\alpha_{g,pl}$ value of CTOA was determined. These properties are related by

$$\alpha_{g,pl} = \kappa \alpha_t \qquad [37]$$

where κ is a fitting parameter. By assessing α_t it is possible to evaluate the energy criterion for elastic-plastic fracture.

In Yoshida and Hallett (2008) they equilibrated soils to a range of water potentials before fracture testing. One sample was dried to −50 kPa and then wetted back to −5kPa. CTOA was related to the most negative water potential experienced by the soil (Fig. 5–8), but as rewetting had minimal impact they demonstrated the importance of particle rearrangement and consolidation caused by shrinkage stresses. With knowledge of the Young's Modulus of the soil, CTOA can be used to evaluate the energy criteria for fracture as described in Eq. [32]

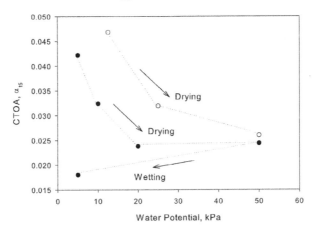

Fig. 5–8. Influence of matric potential on the crack tip opening angle, CTOA, measured 5 mm from the crack tip, α_{t5}. The open symbols are for a paddy soil and the closed symbols for pure montmorillonite. Upon rewetting of a specimen at −50 to −5 kPa, a drop in CTOA is observed. Drawn from data presented in Yoshida and Hallett (2008).

(Kolednik and Turner, 1994). It offers considerable potential to understanding crack propagation in soil.

Toward an Understanding of Soil Structure Dynamics

Earlier in this chapter, the importance of soil crack propagation in the understanding of soil structure dynamics was described. So far the work reviewed has focused primarily on the fracture mechanisms of individual cracks, usually with an external mechanical stress applied in laboratory tests. Ahuja et al. (2006) identified soil structure dynamics, particularly through the growth of cracks, as a major area requiring greater effort in future research on water transport. Few studies have tackled both water flow and crack dynamics in the same study, with the research by Vogel et al. (2005) presenting the only study we could find that incorporates fracture mechanics.

Another hydrological driver of crack propagation are cycles of wetting and drying. Various studies demonstrate cracking processes in drying soils (Yoshida and Adachi, 2001) and have described the processes using fracture mechanics (Konrad and Ayad, 1997; Morris, 1992). Other research has identified direct links between water potential, water content and fracture properties (Mullins and Panayiotopoulos, 1984; Snyder and Miller, 1985; Yoshida and Hallett, 2008). The volume of research conducted in this area, however, has been too small for major advances to have been achieved.

The shrinkage behavior of soil has been the subject of a greater volume of research (Groenevelt and Grant, 2004; Peng et al., 2009), but only a few studies try to incorporate the interaction between shrinkage and swelling behavior of soil and fracture mechanisms (Chertkov, 2002;Chertkov, 2003). Crack networks are highly complex in soil so statistical physics may be necessary to consider how cracks initiate, branch and interact. Nevertheless, this approach requires a fundamental understanding of the propagation of individual cracks, as described as the central focus of this chapter.

Conclusions

Soil structure, by definition, is about complexity. This brings a significant challenge in understanding how soil structure forms, changes over time and influences a myriad of processes ranging from microbial habitats to water transport. This chapter has focused on a dominant process in soil structure dynamics, the growth of cracks. Approaches from materials science and rock mechanics provide great opportunities for soil scientists to gain greater understanding of the fracture mechanics underlying crack propagation. So far the research in this area has extended from very simple models, based on the area and strength of water and soil bonds on fracture surfaces, to an incorporation of new fracture mechanics approaches. Linear elastic fracture mechanics was adopted successfully by soil scientists and the approach has shown great promise. However, more recent research using elastic-plastic fracture mechanics provides a better description of the underlying mechanisms.

Including soil structure dynamics in the modeling of water transport is one of the greatest challenges facing soil physicists. By combining research on soil hydrology, crack propagation and shrink/swell dynamics, various studies have started to tackle this significant challenge. However, insufficient research has examined fracture mechanics in soil, so until these processes are understood, the reliability of models and their ability to account for underlying processes will limit how far we can progress.

Acknowledgments

We thank Jonathan Seville, a chemical engineer, for demonstrating the value of bridging disciplines. The James Hutton Institute receives funding from the Scottish Government.

References

Adams, M.J., D. Williams, and J.G. Williams. 1989. The use of linear elastic fracture-mechanics for particulate solids. J. Mater. Sci. 24:1772–1776. doi:10.1007/BF01105704.

Ahuja, L.R., L.W. Ma, and D.J. Timlin. 2006. Trans-disciplinary soil physics research critical to synthesis and modeling of agricultural systems. Soil Sci. Soc. Am. J. 70:311–326. doi:10.2136/sssaj2005.0207.

Aitchison, G.D. 1961. Relationships of moisture stress and effective stress functions in unsaturated soils. In: Proceedings of the conference on pore pressure and suction in soils. Butterworths, London. p. 47–52.

Aitchison, G.D. 1957. The strength of quasi-saturated and unsaturated soils in relation to the pressure deficiency in the pore water. In: Proc. 4th Intl. Conf. Soil Mech. Foundation Eng. 1. Butterworths, London. p. 135–139.

Bishop, A.W. 1959. The principle of effective stress. Teknisk Ukeblad 39:859–863.

Blight, G.E. 1967. Effective stress evaluation for unsaturated soils. J. Soil Mech. Foundations Div. 93:125–148.

Cavagnaro, T.R., L.E. Jackson, J. Six, H. Ferris, S. Goyal, D. Asami, and K.M. Scow. 2006. Arbuscular mycorrhizas, microbial communities, nutrient availability, and soil aggregates in organic tomato production. Plant Soil 282:209–225. doi:10.1007/s11104-005-5847-7.

Chandler, H.W. 1984. The use of non-linear fracture mechanics to study the fracture properties of soils. J. Agric. Eng. Res. 29:321–327. doi:10.1016/0021-8634(84)90087-8.

Chenu, C., and J. Guérif. 1991. Mechanical strength of clay minerals as influenced by an adsorbed polysaccharide. Soil Sci. Soc. Am. J. 55:1076–1080. doi:10.2136/sssaj1991.03615995005500040030x.

Chertkov, V.Y. 2002. Modelling cracking stages of saturated soils as they dry and shrink. Eur. J. Soil Sci. 53:105–118. doi:10.1046/j.1365-2389.2002.00430.x.

Chertkov, V.Y. 2003. Modelling the shrinkage curve of soil clay pastes. Geoderma 112:71–95. doi:10.1016/S0016-7061(02)00297-5.

Chertkov, V.Y., and I. Ravina. 2004. Networks originating from the multiple cracking of different scales in rocks and swelling soils. Int. J. Fract. 128:263–270. doi:10.1023/B:FRAC.0000040989.82613.36.

de Brouwer, J.F.C., K. Wolfstein, G.K. Ruddy, T.E.R. Jones, and L.J. Stal. 2005. Biogenic stabilization of intertidal sediments: The importance of extracellular polymeric substances produced by benthic diatoms. Microb. Ecol. 49:501–512. doi:10.1007/s00248-004-0020-z.

Dexter, A.R. 2004. Soil physical quality—Part I. Theory, effects of soil texture, density, and organic matter, and effects on root growth. Geoderma 120:201–214. doi:10.1016/j.geoderma.2003.09.004.

Ennis, B.J., and G. Sunshine. 1993. On wear as a mechanism of granule attrition. Tribol. Int. 26:319–327. doi:10.1016/0301-679X(93)90068-C.

Farrell, D.A., W.E. Larson, and E.L. Greacen. 1967. A model of the effect of soil variability on tensile strength and fracture. Paper presented at: ASA, CSSA, SSSA annual meetings, Washington, DC. 5–10 Nov. 1967.

Fippin, E.O. 1910. Some causes of soil granulation. Agron. J. 2:106–121. doi:10.2134/agronj1910.00021962000200010022x.

Fisher, R.A. 1926. On the capillary forces in an ideal soil; correction of formulae given by W.B. Haines. J. Agric. Sci. 16:494–503.

Fisher, R.A. 1928. Further note on the capillary forces in an ideal soil. J. Agric. Sci. 18:406–410. doi:10.1017/S0021859600019432.

Gill, W.R., and G.E. Vanden Berg. 1968. Soil dynamics in tillage and traction. Agriculture Handbook No. 316, Agric. Res. Serv., U.S. Dep. of Agric., Washington, D.C.

Griffith, A.A. 1921. The phenomenon of rupture and flow in solids. Philos. Trans. R. Soc. London A 221:163–198. doi:10.1098/rsta.1921.0006.

Griffith, A.A. 1924. The theory of rupture. In: C.B. Biezeno and J.M. Burgers, editors, In: Proc. First Congress Applied Mechanics. Waltman, Delft. p. 55–63.

Groenevelt, P.H., and C.D. Grant. 2004. Analysis of soil shrinkage data. Soil Tillage Res. 79:71–77. doi:10.1016/j.still.2004.03.011.

Groenevelt, P.H., and B.D. Kay. 1981. On pressure distribution and effective stress in unsaturated soils. Can. J. Soil Sci. 61:431–443. doi:10.4141/cjss81-047.

Guérif, J. 1988. Determination of the tensile-strength of soil aggregates—Review and proposal for an improved technique. Agronomie 8:281–288. doi:10.1051/agro:19880401.

Guérif, J. 1990. Factors influencing compaction-induced increases in soil strength. Soil Tillage Res. 16:167–178. doi:10.1016/0167-1987(90)90028-C.

Haines, W.B. 1925. Studies in the physical properties of soils. I. Mechanical properties concerned in cultivation. J. Agric. Sci. 15:178–200. doi:10.1017/S0021859600005669.

Haines, W.B. 1927. Studies in the physical properties of soil. IV. A further contribution to the theory of capillary phenomenon in soil. J. Agric. Sci. 17:264–290. doi:10.1017/S0021859600018499.

Hallett, P.D., A.R. Dexter, and J.P.K. Seville. 1995. The application of fracture mechanics to crack propagation in dry soil. Eur. J. Soil Sci. 46:591–599. doi:10.1111/j.1365-2389.1995.tb01355.x.

Hallett, P.D., and T.A. Newson. 2001. A simple fracture mechanics approach for assessing ductile crack growth in soil. Soil Sci. Soc. Am. J. 65:1083–1088. doi:10.2136/sssaj2001.6541083x.

Hallett, P.D., and T.A. Newson. 2005. Describing soil crack formation using elastic-plastic fracture mechanics. Eur. J. Soil Sci. 56:31–38. doi:10.1111/j.1365-2389.2004.00652.x.

Harison, J.A., B.O. Hardin, and K. Mahboub. 1994. Fracture-toughness of compacted cohesive soils using ring test. J. Geotech. Eng. 120:872–891. doi:10.1061/(ASCE)0733-9410(1994)120:5(872).

Horn, R., H. Taubner, M. Wuttke, and T. Baumgartl. 1994. Soil physical-properties related to soil-structure. Soil Tillage Res. 30:187–216. doi:10.1016/0167-1987(94)90005-1.

Ingles, O.G. 1963. The shatter test as an index of strength for soil aggregates. In: C.J. Osborn, editor, Proc. Tewksbury Symposium on Fracture, University of Melbourne, Melbourne, Australia. p. 284–303.

Ingles, O.G., and S. Frydman. 1963. An examination of some methods for strength measurement in soils. In: Proc. Fourth Australia-New Zealand conference on soil mechanics and foundation engineering. Inst. of Eng., Adelaide, Australia. p. 213–219.

Ingles, O.G., and D. Lafeber. 1966. The influence of volume defects on the strength and strength isotropy of stabilized clays. Eng. Geol. 1:305–310. doi:10.1016/0013-7952(66)90012-3.

Ingles, O.G., and D. Lafeber. 1967. The inititiation and development of crack and joint systems in granular masses. In: Proc. Symposium on stress and failure around underground openings, Dep. of Mining, Univ. of Sydney, Sydney, Australia, Paper No. 7, 52 pp.

Inglis, C.E. 1913. Stresses in a plate due to the presence of cracks and sharp corners. Trans. Inst. Naval Architects 55:219–230.

Irwin, G.R. 1958. Fracture. In: S. Flügge, editor, Handbuch der Physik. Springer-Verlag, Berlin.

Jaeger, F., S. Bowe, H. Van As, and G.E. Schaumann. 2009. Evaluation of 1H NMR relaxometry for the assessment of pore-size distribution in soil samples. Eur. J. Soil Sci. 60:1052–1064. doi:10.1111/j.1365-2389.2009.01192.x.

Keen, B.A. 1924. On the moisture relationships in an ideal soil. J. Agric. Sci. 14:170–177. doi:10.1017/S0021859600003373.

Kirkham, D., M.F. DeBoodt, and L. DeLeenheer. 1959. Modulus of rupture determination of undisturbed soil cores. Soil Sci. 87:141–144. doi:10.1097/00010694-195903000-00004.

Kolednik, O., and C.E. Turner. 1994. Application of energy-dissipation rate arguments to ductile instability. Fatigue Fracture Eng. Mater. Struct. 17:1129–1145. doi:10.1111/j.1460-2695.1994.tb01403.x.

Konrad, J.M., and R. Ayad. 1997. Desiccation of a sensitive clay: Field experimental observations. Can. Geotech. J. 34:929–942.

Kubiëna, W.L. 1938. Micropedology. Collegiate Press, Ames, IA. 243 pp.

Lachenbruch, A.H. 1961. Depth and spacing of tension cracks. J. Geophys. Res. 66:4273–4292. doi:10.1029/JZ066i012p04273.

Lachenbruch, A.H. 1962. Mechanics of thermal contraction cracks and ice-wedge polygons in permafrost. Geol. Soc. Am. Spec. Pap. 70:69 pp.

Lawn, B.R. 1993. Fracture of brittle solids. 2nd ed. Cambridge Univ. Press, Cambridge, U.K.

Lian, G., C. Thornton, and M.J. Adams. 1993. A theoretical study of the liquid bridge forces between two rigid spherical bodies. J. Colloid Interface Sci. 161:138–147. doi:10.1006/jcis.1993.1452.

Lima, L.A., and M.E. Grismer. 1994. Application of fracture-mechanics to cracking of saline soils. Soil Sci. 158:86–96. doi:10.1097/00010694-199408000-00002.

Loades, K.W., A.G. Bengough, M.F. Bransby, and P.D. Hallett. 2010. Planting density influence on fibrous root reinforcement of soils. Ecol. Eng. 36:276–284. doi:10.1016/j.ecoleng.2009.02.005.

Markgraf, W., R. Horn, and S. Peth. 2006. An approach to rheometry in soil mechanics- Structural changes in bentonite, clayey and silty soils. Soil Tillage Res. 91:1–14. doi:10.1016/j.still.2006.01.007.

McClintock, F.A., and J. Walsh. 1962. Friction of Griffith cracks in rocks under pressure. Proc. 4th. U.S. Natl. Cong. Appl. Mech. 2:1015–1019.

Mickovski, S.B., P.D. Hallett, M.F. Bransby, M.C.R. Davies, R. Sonnenberg, and A.G. Bengough. 2009. Mechanical reinforcement of soil by willow roots: Impacts of root properties and root failure mechanism. Soil Sci. Soc. Am. J. 73:1276–1285. doi:10.2136/sssaj2008.0172.

Morris, P.H. 1992. Cracking in drying soils. Can. Geotech. J. 29:263–277. doi:10.1139/t92-030.

Mullier, M.A., J.P.K. Seville, and M.J. Adams. 1987. A fracture-mechanics approach to the breakage of particle agglomerates. Chem. Eng. Sci. 42:667–677. doi:10.1016/0009-2509(87)80027-1.

Mullins, C.E., P.S. Blackwell, and J.M. Tisdall. 1992. Strength development during drying of a cultivated, flood-irrigated hardsetting soil. 1. Comparison with a structurally stable soil. Soil Tillage Res. 25:113–128. doi:10.1016/0167-1987(92)90106-L.

Mullins, C.E., and K.P. Panayiotopoulos. 1984. The strength of unsaturated mixtures of sand and kaolin and the concept of effective stress. J. Soil Sci. 35:459–468. doi:10.1111/j.1365-2389.1984.tb00303.x.

Murdoch, L.C. 1993a. Hydraulic fracturing of soil during laboratory experiments. 1. Methods and observations. Geotechnique 43:255–265. doi:10.1680/geot.1993.43.2.255.

Murdoch, L.C. 1993b. Hydraulic fracturing of soil during laboratory experiments. 2. Propagation. Geotechnique 43:267–276. doi:10.1680/geot.1993.43.2.267.

Nearing, M.A. 1995. Compressive strength for an aggregated and partiality saturated soil. Soil Sci. Soc. Am. J. 59:35–38. doi:10.2136/sssaj1995.03615995005900010005x.

Newman, J.C., M.A. James, and U. Zerbst. 2003. A review of the CTOA/CTOD fracture criterion. Eng. Fract. Mech. 70:371–385. doi:10.1016/S0013-7944(02)00125-X.

Nichols, J.R., and M.E. Grismer. 1997. Measurement of fracture mechanics parameters in silty-clay soils. Soil Sci. 162:309–322. doi:10.1097/00010694-199705000-00001.

Nieber, J.L. 1981. Simulation of fracturing of a desiccating soil: Stress analysis, Winter Meeting. ASAE, Chicago, IL. Paper No. 81–2512, 28 pp.

Niwa, Y., and S. Kobayashi. 1967. Failure criterion of cement mortar under triaxial compression. Memoirs Faculty Eng. Kyota Univ. 1(1):1–15.

Or, D. 1996. Wetting-induced soil structural changes: The theory of liquid phase sintering. Water Resour. Res. 32:3041–3049. doi:10.1029/96WR02279.

Pande, G.N., and S. Pietruszczak. 1990. A rational interpretation of pore pressure parameters. Geotechnique 40:275–279. doi:10.1680/geot.1990.40.2.275.

Peng, X., J. Dorner, Y. Zhao, and R. Horn. 2009. Shrinkage behaviour of transiently- and constantly-loaded soils and its consequences for soil moisture release. Eur. J. Soil Sci. 60:681–694. doi:10.1111/j.1365-2389.2009.01147.x.

Peth, S., J. Nellesen, G. Fischer, and R. Horn. 2010. Non-invasive 3D analysis of local soil deformation under mechanical and hydraulic stresses by µCT and digital image correlation. Soil Tillage Res. 111:3–18. doi:10.1016/j.still.2010.02.007.

Raats, P.A.C. 1984. Mechanics of cracking soils. In: J. Bouma and P.A.C. Raats, editors, ISSS symposium on water and solute movement in heavy clay soils, Wageningen, the Netherlands. p. 23–38.

Rahman, Z.A., D.G. Toll, D. Gallipoli, and M.R. Taha. 2010. Micro-structure and engineering behaviour of weakly bonded soil. Sains Malays. 39:989–997.

Ramirez-Flores, J.C., S.K. Woche, J. Bachmann, M.O. Goebel, and P.D. Hallett. 2008. Comparing capillary rise contact angles of soil aggregates and homogenized soil. Geoderma 146:336–343. doi:10.1016/j.geoderma.2008.05.032.

Richards, L.A. 1953. Modulus of rupture as an index of crusting of soil. Soil Sci. Soc. Am. Proc. 17:321–323. doi:10.2136/sssaj1953.03615995001700040005x.

Roger-Estrade, J., G. Richard, A.R. Dexter, H. Boizard, S. De Tourdonnet, M. Bertrand, and J. Caneill. 2009. Integration of soil structure variations with time and space into models for crop management. A review. Sustainable Agric. 29:135–142. doi:10.1051/agro:2008052.

Russell, E.W. 1938. Soil structure. Imp. Bur. Soil Sci. 37:1–40.

Simons, S.J.R., J.P.K. Seville, and M.J. Adams. 1994. An analysis of the rupture energy of pendular liquid bridges. Chem. Eng. Sci. 49:2331–2339. doi:10.1016/0009-2509(94)E0050-Z.

Six, J., H. Bossuyt, S. Degryze, and K. Denef. 2004. A history of research on the link between (micro)aggregates, soil biota, and soil organic matter dynamics. Soil Tillage Res. 79:7–31. doi:10.1016/j.still.2004.03.008.

Snyder, V.A. 1980. Theoretical aspects and measurement of tensile strength in unsaturated soils. Ph.D. Thesis. Cornell University, Ithaca, NY.

Snyder, V.A. 1987. Mechanical equilibrium in externally loaded unsaturated granular similar media. Soil Sci. Soc. Am. J. 51:1413–1424. doi:10.2136/sssaj1987.03615995005100060005x.

Snyder, V.A., and R.D. Miller. 1985. Tensile strength of unsaturated soils. Soil Sci. Soc. Am. J. 49:58–65. doi:10.2136/sssaj1985.03615995004900010011x.

Snyder, V.A., and R.D. Miller. 1989. Soil deformation and fracture under tensile forces. In: W.E. Larson, editor, Mechanics and related processes in structured agricultural soils. Kluwer Academic Publ., Dordrecht, the Netherlands. p. 23–35.

Sparks, A.D. 1963. Theoretical considerations of stress equations for partly saturated soils. Proc. 3rd. Regional Conf. Africa Soil Mech. 1, Salisbury, Rhodesia, p. 215–218.

Taina, I.A., R.J. Heck, and T.R. Elliot. 2008. Application of X-ray computed tomography to soil science: A literature review. Can. J. Soil Sci. 88:1–20. doi:10.4141/CJSS06027.

Terzaghi, K. 1943. Theoretical soil mechanics. Chapman and Hall, London.

Turner, C.E., and O. Kolednik. 1997. A simple test method for energy dissipation rate, CTOA and the study of size and transferability effects for large amounts of ductile crack growth. Fatigue Fract. Eng. Mater. Struct. 20:1507–1528. doi:10.1111/j.1460-2695.1997.tb01507.x.

Vogel, H.J., H. Hoffmann, A. Leopold, and K. Roth. 2005. Studies of crack dynamics in clay soil—II. A physically based model for crack formation. Geoderma 125:213–223. doi:10.1016/j.geoderma.2004.07.008.

Vomocil, J.A., L.J. Waldron, and W.J. Chancellor. 1961. Soil tensile strength by centrifugation. Soil Sci. Soc. Am. Proc. 25:176–180. doi:10.2136/sssaj1961.03615995002500030011x.

Warkentin, B.R. 2008. Soil structure: A history from tilth to habitat. Adv. Agron. 97:239–272. doi:10.1016/S0065-2113(07)00006-5.

Williams, J., and C.F. Shaykewich. 1970. The influence of soil water matric potential on the strength properties of unsaturated soil. Soil Sci. Soc. Am. Proc. 34:835–840. doi:10.2136/sssaj1970.0 3615995003400060010x.

Yokobori, T., M. Ohashi, and M. Ichikawa. 1965. The interaction of two collinear asymmetrical elastic cracks. Reports of the research institute for strength and fracture of materials. Tohoku Univ. 1(2):33–39.

Yoshida, S., and K. Adachi. 2001. Effects of cropping and puddling practices on the cracking patterns in paddy fields. Soil Sci. Plant Nutrition 47:519–532. doi:10.1080/00380768.2001.104 08416.

Yoshida, S., and P.D. Hallett. 2008. Impact of hydraulic suction history on crack growth mechanics in soil. Water Resour. Res. 44: W00C01. doi:10.1029/2007WR006055.

Young, I.M., and J.W. Crawford. 2004. Interactions and self-organization in the soil-microbe complex. Science 304:1634–1637. doi:10.1126/science.1097394.

Young, I.M., and C.E. Mullins. 1991. Factors affecting the strength of undisturbed cores from soils with low structural stability. J. Soil Sci. 42:205–217. doi:10.1111/j.1365-2389.1991.tb00402.x.

Zhang, B., P.D. Hallett, and G. Zhang. 2008. Increase in the fracture toughness and bond energy of clay by a root exudate. Eur. J. Soil Sci. 59:855–862. doi:10.1111/j.1365-2389.2008.01045.x.

Chapter 6

Dynamics of Soil Macropore Networks in Response to Hydraulic and Mechanical Stresses Investigated by X-ray Microtomography

Stephan Peth, Jens Nellesen, Gottfried Fischer, Wolfgang Tillmann, and Rainer Horn

Abstract

Soil structure is associated with a complex organization of pore spaces playing a fundamental role for soil functioning by governing multiscale interactions of physical, chemical and biological processes. Studying soil processes and their interaction in structured soils is complicated since soil structure is dynamic leading to temporally variable soil functions depending on soil structure evolution or degradation. Investigating pore space dynamics and its relation to soil functions is a challenging task due to the fact that soil pores are organized in three-dimensional networks and the opaque solid soil constituents prevent direct observations of the pore space. This may be overcome by modern non-invasive techniques such as X-ray computed microtomography (XMCT) allowing a detailed description of the internal modification of pore networks upon changes in boundary conditions. This chapter demonstrates the potential use of XMCT in combination with quantitative image analysis procedures to study the effect of hydrologic/mechanical stresses on the evolution/degradation of soil structure in a loess soil as an example.

Abbreviations: CT, computed tomography; PCSD, pore cluster size distribution; XCT, X-ray computed tomography; XMCT, X-ray computed microtomography.

G. Fischer and J. Nellesen, RIF e.V. Institut für Forschung und Transfer, Joseph-von-Fraunhofer-Str. 20, D-44227 Dortmund, Germany (gottfried.fischer@rif-ev.de; jens.nellesen@rif-ev.de); R. Horn, Christian-Albrechts-Universität zu Kiel, Institute for Plant Nutrition and Soil Science, Olshausenstr. 40, D-24118 Kiel, Germany (rhorn@soils.uni-kiel.de); S. Peth, Universität Kassel, Department of Soil Science, Nordbahnhofstraße 1a, D-37213 Witzenhausen, Germany (peth@uni-kassel.de); W. Tillmann, Technische Universität Dortmund, Faculty of Mechanical Engineering, Institute of Materials Engineering, Leonhard-Euler-Str. 2, D-44227 Dortmund, Germany (wolfgang.tillmann@tu-dortmund.de).

doi:10.2134/advagricsystmodel3.c6

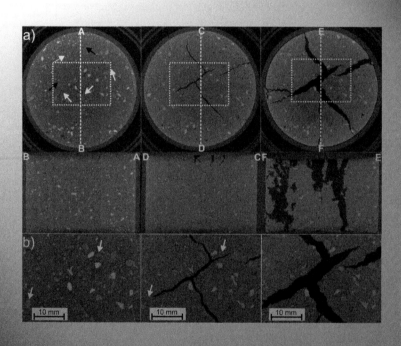

oil functional properties depend to a large extent on soil structure. The geometrical arrangement and morphology of pores, that is, their shape, surface roughness, orientation, size distribution, tortuosity, connectivity and continuity, compose the *infrastructure* for soil organisms and plant roots. Most soils are heterogeneous and structured which increases complexity of the local environment in soil microbial habitats where *abiotic* factors (physical and chemical) control and interact with *biotic* factors (e.g., bioturbation, mineralization, assimilation, exsudation). Some authors use the term "functional architecture" to emphasize the intimate link between soil structure and soil functions which is determined by numerous process interactions (Pierret et al., 2007; Baveye, 2010). Understanding and finally quantifying the complex interactions of physical, chemical, and biological soil functions within soil pore spaces has become an important task for soil scientists to design more environmentally sustainable and efficient cropping systems. It is widely recognized that soil structure plays a fundamental role in those interactions (Young and Crawford, 2004; Horn and Smucker, 2005; Nunan et al., 2006; Pierret et al., 2007; Smucker and Hopmans, 2007) and hence should be incorporated in the development of process-based models which are capable of integrating the physical, chemical and biological heterogeneities naturally occurring in soils (Wu et al., 2004) and which may then serve as a tool for decision making in soil management.

Available process-oriented models in soil science mostly aim at predicting (i) transport of heat, water, solutes (nutrients and pollutants), bacteria, colloidal (clays, Fe and Al oxides, humic materials) and gaseous compounds (e.g., CO_2, N_2O, CH_4) within the soil pore space, (ii) their fluxes between the pedosphere and adjacent spheres (biosphere, hydrosphere, and atmosphere), and (iii) exchange and transformation reactions between solid, liquid, and gas phases. In the last decade significant progress has been made in developing sophisticated models and modeling techniques tackling a range of problems in predicting transport and transformation phenomena in soils (or porous media in general) such as preferential flow (Gerke, 2006), colloidal transport (Bradford and Torkzaban, 2008), root water and nutrient uptake (Hopmans and Bristow, 2002), flow behavior across scales (Lehmann et al., 2008), multiphase flow (Adler, 1995), flow through the rhizosphere (Darrah et al., 2006), reactive transport (Simunek et al., 2008), coupled heat and water flow (Bittelli et al., 2008), and pore scale organic matter decomposition (Monga et al., 2008) to name a few. Despite the accomplished improvements of models major challenges still remain: From our point of view the most pertinent problems in predictive modeling are (i) to fully account for the complexity of natural soil pore systems at the pore scale and its role in habit functioning in relation to soil structural heterogeneities on various scales (Nunan et al., 2006; Peth et al., 2008a) and (ii) to include the dynamic nature of pore spaces and related properties caused by biophysical, hydraulic and mechanical stresses (Gerke, 2006; Smucker and Hopmans, 2007; Horn and Peth, 2009). For example, in most hydraulic models water flow through variably saturated porous media is described by the Richards equation (Richards, 1931), which requires rigidity of the pore space. That is, the functional relationship between hydraulic conductivity and pressure head is considered constant with respect to pore water pressure throughout the modeled time period. In other words swelling and shrinkage is assumed to have no impact on the permeability of the pore system which is conceptually incorrect.

Some fundamental studies on the mechanics of cracking and swelling clay soils seeking relations between volume change and water flow have been conducted in the 1980s by Raats (1984, 1987) and Smiles (1981). Those very interesting approaches provided a theoretical framework for the extension of Richards equation to non-rigid pore systems but had to satisfy material continuity thus neglecting preferential flow along cracks (Smiles 1981, 2000). The formulations are indifferent to three-dimensional volume change of aggregates and cracks. While this was demonstrated to be appropriate for clay suspensions and saturated mine tailings (Smiles 2000), the approach yet has to be extended to multidimensional unsaturated systems with complex structures and its use in agriculture therefore still remains uncertain (Smiles 1995). One of the limitations is that detailed understanding of the involved microscale deformation processes and associated re-arrangement of pores and their effect on transport is missing, which may explain the persistent acceptance of static pore architectures in Richards-based models.

Being dependent on water and gas transport also biogeochemical reactions in natural soils are modified by the dynamics of soil pore spaces and hence calculated fluxes derived from models assuming structural rigidity potentially lead to false predictions. Especially structurally unstable soils such as plow layers evolve due to both post-tillage shrinking and swelling (Or and Ghezzehei, 2002; Peng et al., 2007) and re-compaction after wheeling (Horn et al., 1995; Wiermann et al., 2000). This gives rise to highly dynamic pore space geometries and transport functions which vary within the course of a year. But also subsoils containing reasonable amounts of swelling clay minerals or being subjected to increasing mechanical stresses due to field traffic are characterized by variable pore space architectures. On the other hand it has been demonstrated using X-ray tomography that soil structure dynamics is significantly influenced by macrofauna such as earthworms (Jégou et al., 1999, Schrader et al., 2007, Capowiez et al., 2011) and can even result from microbial activity (Feeney et al., 2006). To the best of our knowledge most of the currently used transport models in soil science do not include the time-dependent dynamics of pore systems and transport functions.

An attempt to model soil structure dynamics on the pore scale based on soil mechanical and rheological properties has been presented by Or and Ghezzehei (2002). The proposed model could predict the geometrical re-arrangement of monosized aggregate spheres from changes in external (mechanical) as well as internal (capillary) stresses during the deformation process and was later upscaled to an aggregate bed scale (Ghezzehei and Or, 2003). Such modeling approaches are important to study the principles of soil structure dynamics and to estimate changes in hydraulic properties induced by mechanical and hydraulic stresses. However, pore space geometry had to be simplified in terms of structural entities (unit sized spherical aggregates) and their spatial arrangement (rhombohedral packing of aggregates evoking homogeneous symmetrical distribution of stresses) for the sake of mathematical tractability. Structure dynamics in natural soils, however, is much more complex and studies focusing on microscale deformation phenomena are still limited (Gerke, 2006).

Investigating pore space dynamics and its relation to soil functions is a challenging task since soil pores are organized in three-dimensional networks and the opaque solid soil constituents prevent direct observations of the pore space. On the other hand traditional methods employed in soil structure analysis are limited because they are either restricted to two dimensions and involve destructive sample preparation (thin sections), or because the method itself induces soil deformation by modifying the internal hydraulic stress state evoking swelling and shrinking of the sample (e.g., pore size distribution derived from water retention functions). While swelling and shrinking processes can be studied during wetting and drying by measuring bulk soil volume changes as shown by Peng et al. (2007) and to a certain degree also by quantifying crack volume changes by digital image analysis of soil surface images, Peng et al. (2006) detailed three-dimensional information of internal pore geometry changes is not obtained by those methods. X-ray computed tomography (XCT) as a nondestructive technique is not subject to these limitations thus providing an almost ideal tool to track pore space modifications with changes in stress conditions. Since the XCT scanners have significantly been improved in the last decade in terms of spatial resolution (down to $1~\mu m^3$ XMCT) pore size modifications in soil in the range from a few millimeters to micrometers can be studied. Some examples showing the potential but also the limitations of XMCT in research on soils and sediments are provided in Peth (2010).

Once a three-dimensional digital image dataset has been reconstructed from a scanned sample it portrays a snapshot of the local arrangement of solids and voids in space under the prevailing environmental conditions. Changing the boundary condition (e.g., mechanical or hydraulic stress state) potentially induces modifications of the spatial arrangement of pores and solids. The associated structural evolution of a soil under various boundary conditions can be documented by repeated tomographic imaging of the sample and comparing the initial or a previous state of the pore system with a modified state. Finally, since data is available in digital form, it is possible to visualize and quantify observed changes by means of data processing and image analysis methods. The goal of this chapter is to demonstrate the potential use of XMCT in combination with quantitative image analysis procedures to study the effect of mechanical/hydrological stresses on the evolution of soil structure in a loess soil as an example. Part of the results presented here has been published in a previous study (Peth et al., 2010) but will be complemented by the aspect how water influences the deformation behavior when mechanical stresses are applied. Microscale local deformation phenomena and associated changes in the pore network resulting from purely hydraulic, purely mechanical, and coupled stress conditions will be shown and discussed in view of their potential to better describe the non-rigidity of porous media.

Materials and Methods

Materials and Methods have partly been described in a previous paper (Peth et al., 2010), so only a summary of the previous methods will be repeated here. The new approach and analysis will be discussed in detail.

Material and Sample Preparation

For this study we used Avdat loess (17% sand, 64% silt, 19% clay) from the Negev desert/Israel which was homogenized and filled air dry into a plastic cylinder of 5.5 cm diameter and 4 cm height at an initial bulk density of 1.2 g cm^{-3}. After subjecting the sample to various hydraulic stresses to develop soil structure by crack formation we applied mechanical stresses at two moisture conditions (air dry and −6 kPa) to analyze the deformation behavior using XMCT.

Initial Structure Formation by Swelling–Shrinking

To investigate the effect of hydraulic stresses on initial soil structure formation in a homogenized soil we have wetted the sample by capillary saturation and subsequently air dried it at room temperature. The sample was scanned in three states:

(i) initially homogenized air dry condition (hom); (ii) after wetting to near saturation (sat); and (iii) after subsequent drying (dry).

Mechanical Loading Test

To study the stability of the structured (dry) sample to mechanical stresses we
have applied stepwise increasing mechanical loads with nominal stresses of 10,
20, 30, 40, 50, 60, 80, 100, 120, 150, 200, 300, and 400 kPa. After shrinkage a micro
topography was formed at the sample surface. Therefore, we decided to use force
as a loading parameter instead of stress. Loading steps were then accordingly:
24, 47, 71, 95, 119, 142, 190, 237, 285, 356, 475, 712, and 950 N. After a load of 950 N
in dry condition, however, almost full contact between the loading plate and the
sample was established. At selected loading steps (0 = initially unloaded, 47, 119,
237, 475, and 950 N) we have scanned the sample immediately after unloading
when rebound movements had ceased.

After soil compression was completed for the sample in air dry condition we
have slowly rewetted the soil (subsequently referred to as wet) from the bottom
of the sample on a special suction device to a matric potential of −6 kPa (Fig. 6–1).
Wetting was done with water at sub-atmospheric pressure (−6 kPa) to ensure that
the previously formed structure would not collapse on complete saturation of the
pore space due to a loss of menisci forces at inter-particle contacts. The suction
plate assembly was designed such that the ceramic plate can directly be connected to the sample holder allowing the sample to be scanned without removing
it from the sample holder (Fig. 6–1). The same procedure of mechanical compression as described above was performed also for the wet sample, while the sample
remained in the sample holder, except that now pore water was present during
loading which was assumed to influence the compaction process.

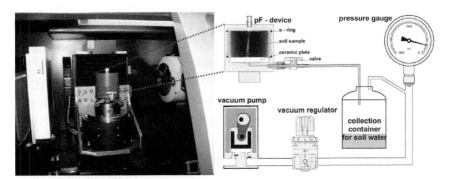

**Fig. 6–1. (right) Special suction plate device adapted for installation in (left) a tomography system
to facilitate multiple scans of a sample at modified boundary conditions (e.g., different hydraulic
stress states or after mechanical loading).**

From the load-displacement curves obtained for all loading steps we determined the pre-compression stress (P_c) as an overall stability indicator for volumetric deformation. Pre-compression stress was determined after Casagrande (1936) using the program RETC (van Genuchten et al., 1991) following a mathematical procedure suggested by Baumgartl and Köck (2004). Pre-compression stress was only determined for the wet sample since for the dry sample the precondition of full contact of the loading plate with the soil was not met.

X-ray Microtomography

Samples were scanned with a *phoenix nanotom* (GE Sensing & Inspection Technologies GmbH, Wunstorf, Germany) at the University of Kiel. Scanning conducted with a maximum X-ray energy of 150 keV and the resulting voxel edge length was 50.6 μm. Reconstruction from the projections was done with the software datos|x (GE Sensing & Inspection Technologies GmbH, Wunstorf, Germany).

Image Processing

Volume rendering and visualization of the three-dimensional pore structures were achieved with Visual Studio Max 2.0 (Volume Graphics GmbH, Heidelberg, Germany). Before further data processing the original reconstructed images were scaled from 32-bit float format down to 8-bit format using the same minimum and maximum grayscale values for all datasets to ensure the same image contrast for all reconstructed volumes.

Segmentation

For the segmentation of the grayscale tomograms we used a local thresholding scheme suggested by Oh and Lindquist (1999), which we have found to best recover soil structure related pore space (Peth et al., 2008a). The method is based on two global threshold values, where voxels below a lower (T_0) and above an upper (T_1) threshold are classified as definitely pore and solid phase, respectively. Voxels with grayscale values in between the two bounding thresholds $(T_0$ and $T_1)$ are associated with a high risk for misclassification and marked in the first step as "uncertain" voxels. In a second step such voxels are classified according to the distribution of voxel grayscale values in their local neighborhood three-dimensionally by means of a geostatistical analysis referred to as *indicator kriging*. During this kriging step the probability for exceeding or falling below the corresponding threshold values is calculated and used as a decision criteria for classification of those voxels associated with high uncertainty (Oh and Lindquist, 1999).

Indicator kriging based thresholding of the grayscale images has been conducted with the software 3DMA-Rock (Lindquist et al., 2005). Currently there is no straight forward procedure to automatically select suitable values for T_0 and T_1. This makes visual inspection of the thresholded binary image in comparison to the original grayscale tomogram an inevitable step to assess whether pores are segmented adequately. If results are not satisfying thresholds are usually manually adapted. On the other hand in doing so bias may be introduced as slight changes in the threshold values potentially correspond to additions or losses of pores. Especially, smaller pores in the soil matrix may be affected by this procedure while the larger structural pores are less sensitive. However, it can still lead to a poor reproducibility of the segmentation results, particularly for the smaller pore size fractions. To circumvent such user related bias by only relying on visual calibration of thresholds we have adapted a pragmatic segmentation procedure based on parameters of a Gaussian fit to the frequency distribution of grayscale values within the sample. After some tests we found that fitting a Gauss curve to the solid peak of the grayscale histogram and calculating the threshold values from the center (α) and standard deviation (σ) of the curve with $T_0 = \alpha - 3\sigma$ and $T_1 = \alpha - 2\sigma$ gave reasonably accurate results in terms of recovery of the pore space while being reproducible at the same time (Fig. 6–2). Assuming that the densities of the solid soil constituents as well as noise and partial volume effects due to solid–void interfaces within the soil matrix are normally distributed, we can state that with a certainty of about 99.7% no solid phase is classified as void by setting T_0 according to $T_0 = \alpha - 3\sigma$. Since the shape of the solid peak is rather symmetric the assumption of normal distribution seems to be reasonable (Fig. 6–2). For the upper limit T_1, above which voxels are classified as solid phase (meaning matrix including unresolved pores), there is a higher risk that pores are incorrectly classified in the binary image as solid phase. However, since such potential pores are usually very small and in any case not all pore space can be resolved in XMCT scans we accepted to cut a number of smaller pores to increase confidence in the statistical results of the finally quantified structure related (secondary) pore space.

Pore Cluster Size Distribution

In a previous study we quantified pore space morphologies based on a medial axis and throat computation approach (Peth et al., 2008a). The method proved to be suitable for pore spaces characterized by spherical/blob like nodal pores which are connected via narrow throat channels or lenticular and elongated pores with a round shaped cross-section (e.g., intra-aggregate biopores). The calculation of a medial axis in a soil with planar like crack surfaces, as it was encountered in this study, produced numerous branching medial axis paths to a large extent resulting from the roughness of the crack surface. In this study we decided therefore to follow a more rigorous description of the pore space in terms of connected

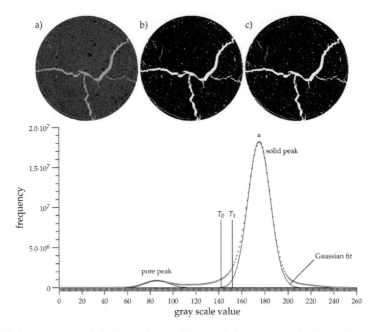

Fig. 6–2. (bottom) Grayscale histogram (green dotted line) of the wet sample at 0 kPa loading condition showing (right) a distinct solid (porous matrix) peak and (left) a less distinct pore peak. The Gaussian fit to the solid peak is shown as red line and corresponding calculated threshold values for pore (<T_0) and solid (>T_1) classification are indicated by vertical black lines. Between T_0 and T_1 voxel classification is based on indicator kriging. (top) Cross-sections through the sample showing (a) the original grayscale tomogram, (b) resulting binary image after the first sweep of the segmentation algorithm and uncertain voxels in red, and (c) the finally segmented (binary) image where white represents the pore space and black the porous matrix, respectively. Voxels associated with high uncertainty in classification which have finally been converted to solid are shown in (c) as red colors. Yellow colors (hardly visible in this image) correspond to "uncertain" voxels which have finally been converted to pore phase after the kriging step.

pore cluster sizes. A pore cluster is defined as a set of connected voids irrespective of their morphology. Cluster sizes are calculated from the number of void voxels multiplied by the voxel unit volume. Note that connectivity is ultimately a question of voxel resolution obtained in a particular scan. Theoretically 100% connectivity is given in the case where all pores (excluding pores within the grain phase itself) would be resolved. However, since reconstructed voxel resolution for the data presented here was the same (50.6 μm) for all scans comparability of results is warranted. Pore cluster size distributions (PCSD) were obtained by calculating all disconnected volumes in the binary images assuming 26-connectivity (i.e., neighboring void voxels are considered to be connected over the 6 faces, 12 edges and 8 corners) using the disconnected pore volume algorithm in 3DMA-Rock. With the given voxel resolution of 50.6 μm the detected pore space is associated with structural pores, which in terms of functional relationships characterizes the fraction of the pore space which is easily accessible and most relevant for gas flux (aeration) and near saturated water transport.

Solid–Void Interface Surface Area

Solid–void interface surface area has been calculated by triangulating the pore wall with a marching cube algorithm (Lorensen and Cline, 1987; Bloomenthal, 1988) implemented in 3DMA-Rock. When calculating surface areas we distinguished between inter-aggregate macropore surface area corresponding to the largest connected pore volume (referred to as *crack surface area*) and intra-aggregate macropore surface area (referred to as *aggregate surface area*) calculated from difference between *total surface area* and *crack surface area*. The two numbers can be interpreted as surfaces which are more (crack surface) and less (aggregate surface) accessible.

Analysis of Strain Localization by X-ray Microtomography

To quantify deformation resulting from hydraulic and mechanical stresses we applied a digital three-dimensional image correlation technique in which gray level gradients in the tomograms caused by attenuation contrast between the above mentioned components of the soil are utilized (Crostack et al., 2008).

In Fig. 6–3 the algorithm of this analysis technique is outlined in which two tomograms are locally matched: To this end, the reference tomogram is divided into cuboid regions which are regularly distributed on a rectangular grid. Each cuboid represents the local microstructure in the reference state. The edge length

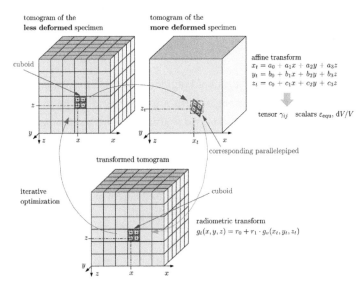

Fig. 6–3. Schematic of the algorithm for the three-dimensional localized strain analysis (from Crostack et al. 2008, slightly modified). Published with permission from Elsevier (Peth et al., 2010).

of the cuboid (mostly cube-shaped) regions is adapted to more or less three times the characteristic microstructural length which is defined as the mean distance between the centers of microstructural objects (like particles, pores). In so doing it is guaranteed that enough voxels with elevated gray level gradients are available in the cuboid regions. In most cases, adjacent cuboidal regions overlap, that is, that the grid spacing is set smaller than the cuboid size.

Each cuboid in the reference tomogram is iteratively mapped onto the corresponding more or less deformed microstructural parallelepiped-shaped region in the comparison tomogram. During this iteration procedure an affine transform

$$
\begin{aligned}
x_t &= a_0 + a_1 x + a_2 y + a_3 z \\
y_t &= b_0 + b_1 x + b_2 y + b_3 z \\
z_t &= c_0 + c_1 x + c_2 y + c_3 z
\end{aligned}
\tag{1}
$$

and a radiometric transform

$$
g_t(x, y, z) = r_0 + r_1 g_v(x_t, y_t, z_t)
\tag{2}
$$

are optimized with the least squares method. From the affine transform (Eq. [1]) the mean deformation gradient tensor $\mathbf{F}$

$$
\mathbf{F} = \begin{pmatrix} a_1 & a_2 & a_3 \\ b_1 & b_2 & b_3 \\ c_1 & c_2 & c_3 \end{pmatrix}
\tag{3}
$$

and subsequently the Lagrangian strain tensor γ (also known as the Green–St. Venant tensor)

$$
\gamma_{ij} = \frac{1}{2} \left(F_{ki} F_{kj} - \delta_{ij} \right)
\tag{4}
$$

are computed.

The affine transforms of *all* cuboid regions were used to calculate *fields* of the displacement vector, the Lagrangian strain tensor γ and of the volume change dV/V sampled on a discrete regular grid within the analyzed tomogram volume.

The deformation gradient tensor $\mathbf{F}$ can be decomposed uniquely into a rotation tensor $\mathbf{R}$ and the positive-semidefinite right stretch tensor $\mathbf{U}$: $\mathbf{F} = \mathbf{RU}$. The scalar quantities, volume change dV/V and equivalent strain ε_{equ} can be derived from deformation gradient tensor $\mathbf{F}$ (Eq. [5]) and eigenvalues (principal values) γ_1, γ_2, and γ_3 of the Lagrangian strain tensor γ (Eq. [6]):

$$
dV/V = \det \mathbf{F} - 1
\tag{5}
$$

$$
\varepsilon_{equ} = \frac{1}{3} \sqrt{(\gamma_1 - \gamma_2)^2 + (\gamma_1 - \gamma_3)^2 + (\gamma_2 - \gamma_3)^2}
\tag{6}
$$

Results and Discussion

Soil Structure Formation by Hydraulic Stresses

The modification of soil structure by internal hydraulic stresses has been shown, using the same material, for three different initial bulk densities in Peth et al. (2010). Here we present data on the sample with an intermediate bulk density of 1.2 g cm^{-3} which was subsequently mechanically loaded under dry and wet condition to follow the dynamics of the pore space with changes in stress state variables. Since the soil was filled into the sample holder in air dry homogenized condition the inter-particle strength was initially very low. This has resulted in the generation of surface discontinuities (micro-fissures) during sample handling at the beginning of the experiment. It turned out that the initial discontinuities served as nuclei for crack development subsequently influencing structure evolution resulting from hydraulic stresses. However, such discontinuities are likely to occur also under natural conditions where they are known to influence the stress field and hence crack initiation (Weinberger, 1999; Yoshida and Adachi, 2004) and therefore are not considered to introduce bias in the outcome of this study.

After wetting the previously dry sample from the bottom we observed the generation of numerous new macro-pores and a widening and downward propagation of the initial micro-cracks (Fig. 6–4a). Looking into detail we further note that the deformation processes resulting from the capillary forces of the infiltrating water lead simultaneously to the generation of new pores while in other places initially available pores/cracks were closed indicating a heterogeneous distribution of internal hydraulic stresses. Opening and closing cracks are shown in Fig. 6–4 by white and black arrows, respectively. Dexter (1991), who described a change in pore morphology in response to fast wetting of soil, pointed out a combined effect of differential swelling and pressure build-up in entrapped air causing mechanical failure of the soil. In extreme cases this failure could result in complete soil slacking into separate micro-aggregates and at lower wetting rates in the formation of arrays of micro-cracks making the soil more friable (Dexter et al., 1984). Apart from the crack in our case more spherical type of pores were formed in the soil matrix and as will be seen later these pores were remarkable stable presumably owing to the influence of the pore-shape on stress distribution (less stress concentration at pore edges). We attribute the formation of macropores on wetting to the heterogeneous distribution of hydraulic stresses resulting from the differential propagation of the wetting front through the soil caused by local differences in capillarity and sorptivity. In this case both menisci forces pulling particles into water filled pore space in combination with air pressure build-up in encapsulated voids may have caused the generation of the observed

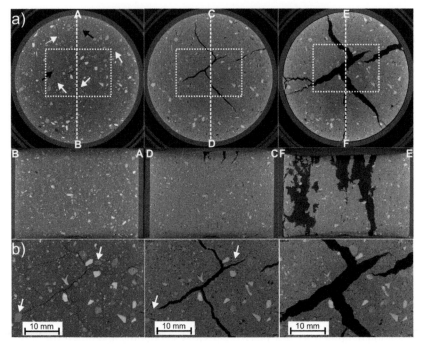

Fig. 6–4. (a) Cross-sectional and longitudinal-section tomogram view of the loess sample representing the change in soil structure after wetting and drying. (left) homogenized air dry condition
(hom); (middle) after wetting to near saturation (sat); (right) after subsequent drying (dry). A
detailed view of the rectangular area (white dashed square) is shown in (b). White and black
arrows show cracks which opened or closed following changes in hydraulic stress, respectively.

spherical and lenticular pores. This effect is assumed to be supported by the weak
structure and low interparticle strength of the homogenized soil but has also
been observed in a previous study in natural soil aggregates (Peth et al., 2008b).
Subsequent air drying of the sample widely opened the initial cracks and generated new secondary order cracks. The large cracks almost completely penetrated
the sample creating a large connected crack volume (Fig. 6–5). The rearrangement
of the soil particles by capillary forces during wetting strongly influenced the pore
size distribution. More than 150,000 new smaller macro-pores were formed indicated by the steeper frequency distribution of pore volumes (Fig. 6–5b). Although
the number of intra-aggregate macro-pores was reduced by $\Delta f > 17{,}000$ following air drying the pore size distribution remained remarkably stable compared
to the change resulting from wetting (Fig. 6–5c). This suggests that the deformation within the newly formed aggregates accommodating the gain of crack
volume must have been compensated by the reduction of pores $<10^{-3}$ mm^3, that
is, pores below the resolution limit. Based on the volume fractions of the largest
connected pore cluster (associated with the crack) and the remaining macro-

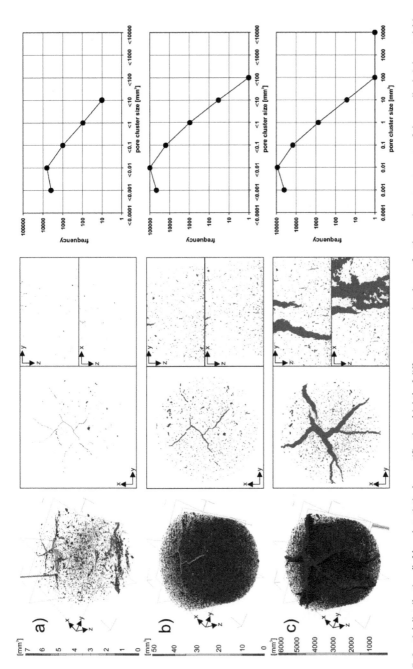

Fig. 6–5. Spatial distribution (left) and statistical quantification (right) of different pore size classes for the various treatments of the swell–shrink test: (a) homogenized air dry condition (hom); (b) after wetting to near saturation (sat) and (c) after subsequent drying (dry). Note that ranges for the color codes have been adjusted to optimally visualize the various pore size classes for each treatment. The decrease in frequency for the smallest class (<0.001 mm³) is related to the detection limit which was equal to 0.0005 mm³. Published with permission from Elsevier (Peth et al., 2010).

pore volume (associated with intra-aggregate macro-pores) in the sample (Table 1) we calculated that the increase in crack volume was only to a small degree (1.9%) accommodated by a reduction of intra-aggregate macro-pores. We assume that hydraulic stresses (menisci forces) exceeded the inter-particle strength only at higher matric potential values where macro-pores are air filled and thus not affected by capillary contraction.

Local deformation analysis by digital image correlation revealed a heterogeneous distribution of volume strain (dV/V) in both vertical and horizontal direction (Fig. 6–6).

During the mechano-hydraulic loading of samples a superposition of pore growth, crack opening, relative displacement between aggregates and real deformation takes place, which the digitial image correlation algorithm does not distinguish. Therefore, apparently elevated strain values were found especially in regions of cracks between aggregates (Fig. 6–3).

If the crack opening (the change of the gap between the aggregates) from the reference state to the comparison state is relatively small within a cuboidal analysis region, the iterative algorithm is likely to converge. In the opposite case, the algorithm is likely to diverge.

To avoid divergence of the algorithm or unrealistic strain values cuboidal analysis regions were excluded from the analysis by the following way: The macro pore space (i.e., the initiated and propagated macro cracks) and the surroundings were segmented by interactive three-dimensional region growing. During region growing contiguous voxels below a preassigned global threshold were marked as "bad voxels." Cuboidal regions which contain too many bad voxels given also by a threshold number were excluded from the analysis.

As seen in Fig. 6–6 deformation was more prominent at the top and in the center of the sample. The patchy pattern of volume strain reflects the heterogeneity of mechanical properties and probably also stress distribution. Deformation was further observed to be anisotropic indicated by the flattened shape of the Lagrangian strain tensor γ (Fig. 6–7). Note that strain tensors are represented by cuboidal glyphs in Fig. 6–7 which are orientated according to the eigenvectors and scaled according to the unsigned eigenvalues. Isotropic deformation (equal eigenvalues) means that a cuboidal glyph corresponds to a cube with edge lengths $a = b = c$ (in terminology of ellipsoids this would be a sphere). The more anisotropic the deformation the larger is the ratio between the longest and shortest axis of the cuboids ($a > b > c$). The strain tensor glyphs become inclined into the vertical direction toward the crack surface where also strain values increase shown by the blue color (Fig. 6–7). This corresponds to the macroscopic observation that shrinkage was strongest at the junction of cracks associated with a more pronounced vertical settlement

Table 1. Pore volume redistribution due to changes in hydraulic stress state (hom = homogeneous initially dry; sat = saturated; dry = drying). The largest connected pore cluster is referred to as crack.

Sample	Crack volume	Intra-aggregate macro-pore volume
	———————— mm³ ————————	
hom	6.5	91.9
sat	46.6	1054.5
dry	5786.1	944.9

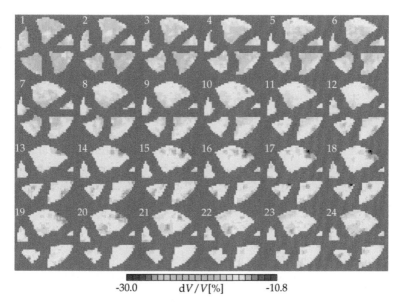

Fig. 6–6. Spatial distribution of dV/V within the newly formed aggregates during the swell–shrink test in different layers. Image correlation analysis was conducted for the sample pair sat vs. dry. Layer 1 is situated approximately 3–4 mm from the sample surface. Layer 24 is the bottom layer. Published with permission from Elsevier (Peth et al., 2010).

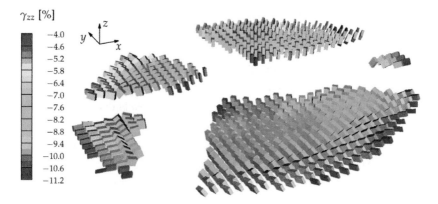

Fig. 6–7. Spatial orientation of the Lagrangian strain tensor γ represented as cuboids in a horizontal cross-section through the specimen. The cuboids are colored according to the tensor entry γ_{zz}. Published with permission from Elsevier (Peth et al., 2010).

of the sample in the middle part (Fig. 6–4a). We assume that while the crack opened due to horizontal shrinkage the drying process in the middle of the sample was enhanced by venting the crack. In consequence, the surface area for evaporation at the aggregate-crack interface will be enlarged. Ritchie and Adams (1974) suggested that air convection currents develop in cracks when they exceed an opening of 4–5 mm. Depending on drying rates this effect will lead to a more (high drying rates) or less (low drying rates) inhomogeneous distribution of tensile stresses and hence capillary contraction which is likely to influence the crack pattern and local volume changes (Yoshida and Adachi, 2004). Also feedback mechanisms between structure related transport processes through the pore network and shrinkage induced pore space modifications are important parameters determining structure dynamics. Vogel et al. (2005) have developed a model simulating crack dynamics based on a lattice of Hookean springs of finite strength. By increasing the initial variance of critical strain they could demonstrate the prominent effect of initial mechanical heterogeneity on the finally developed crack pattern. Local deformation analysis shown in this study has the potential to characterize the heterogeneity of soil mechanical properties and its change with structure evolution providing a physical basis for the determination of material parameters used in models for simulation crack dynamics.

Soil Structure Deformation by Mechanical Stresses

After initial structure development by shrinkage, which has led to a significant re-distribution of pore space by generating cracks and aggregates containing a number of macropores the sample was mechanically loaded first in dry and subsequently in wet condition (matric potential of –6 kPa). The load-displacement curves clearly indicate the influence of water on the deformation behavior with an overall higher settlement under wet condition (Fig. 6–8). Note that the difference in displacement of the wet and dry sample would have to be expected higher if the sample had no stress history due to the previous loading in dry condition. Previous stress application consolidated the sample making it more resistant to subsequent mechanical stresses. Pre-compression stress P_c for the wet sample was calculated to 122 kPa corresponding to a load of 290 N distributed over the total sample surface. According to consolidation theory most of plastic deformation is assumed to occur at loads higher than P_c within the virgin compression range compared to the recompression range (Fig. 6–8).

Measuring soil structural changes by μCT after subjecting the sample to increasing loads revealed remarkable stability of the cracks for loading under dry condition (Fig. 6–9a). At higher loads (>237 N) shear cracks developed predominantly at the sample rim but also within the newly formed aggregates (white arrow

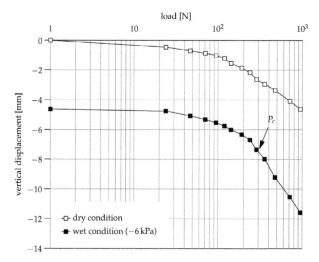

Fig. 6–8. Soil compression curves for the dry and wet condition (matric potential adjusted to −6 kPa). P_c corresponds to the precompression stress.

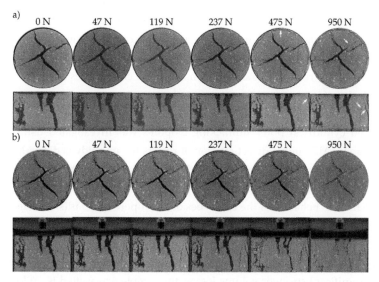

Fig. 6–9. Cross-sectional and longitudinal-section views of tomograms for the different loading steps for the (a) dry condition and (b) wet condition representing the change in soil structure with increasing load. White arrows indicate the development of shear cracks due to loading.

at 950 N in Fig. 6–9a). Nevertheless, except the development of a few shear cracks, soil structure and associated pore size distributions remained rather intact (Fig. 6–9a). This was different when the sample was loaded in wet condition. When loads exceeded the pre-compression stress (corresponding to a force >290 N) a significant change in soil structure was observed mainly associated with the closing of cracks (Fig. 6–9b). Calculations based on the disconnected pore clusters derived from image analysis showed that both total volume fraction of macropores and inter-aggregate crack volume was reduced only little for the dry sample and for the wet sample up to the pre-compression stress but steeply declined for the wet sample at loads higher than P_c (Fig. 6–10). Most of the porosity in the sample is associated with the interconnected crack volume. We calculated a connectivity ratio C of the pore volume of >0.9 ($C = 1$ means that all pores are connected to each other and decreasing C indicates decreasing connectivity) for all tomograms except for the loads of 475 and 950 N of the wet sample where connectivity decreased to 0.89 and finally 0.78, respectively. However, the values show that most of the deformation (volume change) takes place in the interconnected crack and only at high loads of >475 N a significant redistribution of pore cluster sizes also within the aggregates was observed for the wet sample (Fig. 6–11b), while for the dry sample almost no change in pore size distribution could be noticed (Fig. 6–11a). It is recognized that the reduction of inter-aggregate pore volume (crack volume) is associated with a relative increase of intra-aggregate macropores indicated by an upward shift of the relative cluster size distribution function in Fig. 6–11b. Note that the strongest change for the highest load was observed for (i) larger pore cluster sizes which can be explained by a loss of connectivity of the crack generating a number of disconnected crack remnants and (ii) for smaller pore cluster sizes attributed to the compression of intra-aggregate macropores.

In a study by Schäffer et al. (2008) the influence of compression on artificially produced cylindrical pores mimicking biopores was investigated using X-ray tomography. Although a direct relation of results to our study is difficult due to the different morphologies of the cylindrical pores (artificial biopores) and the planar type shrinkage cracks (this study) it is interesting to compare the different mechanical behavior of the two types of macropore systems (shrinkage crack and "biopores") which are both very important in field soils for soil functioning. Similar to our study Schäffer et al. (2008) found a decrease in mean pore diameter of the artificial cylindrical pores by lateral soil displacement during compression with a significant stronger volume loss as the sample was wetted to –6 kPa compared to a more dry state of –30 kPa. However, in contrast to the predominant local soil deformation in the vicinity of cracks in our study they found that the cylindrical macropores were much less affected by compression than the smaller macropores in the rest of the sample (bulk soil). This could be attributed to the influence of macropore morphology (cylindrical vs. planar) and orientation of the pores to the principle directions of stress on the local stress distribution in the sample. Stresses were presumable released by deformation in the bulk soil around the cylindrical macropore with only little local stress concentration at

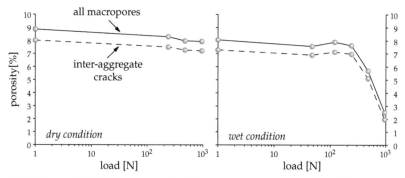

Fig. 6–10. Change in total (all macropores) and inter-aggregate macroporosity with increasing load for the dry condition (left) and wet condition (right).

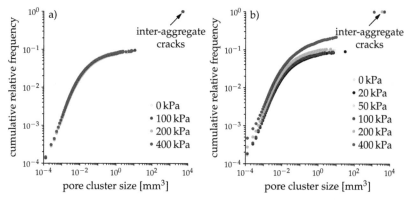

Fig. 6–11. Pore cluster size distribution for the different loading states for the (a) dry condition and (b) wet condition. The points on the top right of each graph represent the inter-aggregate crack volume.

the macropore soil interface. Stress distribution under confined compression can be considered isotropic around the vertical cylindrical macropores leading to an isotropic closure (Schäffer et al., 2008) while we found very inhomogeneous and anisotropic deformation in the case of a crack including both compression and shear deformation finally resulting in a loss of crack connectivity. The non-uniform deformation behavior observed in our study reflects an anisotropic stress distribution which is influenced by the presence of the crack. The heterogeneity and anisotropy of strains is also demonstrated by local strain analysis explained further below. The more complex pore morphology of the crack in this study resulted in a different deformation behavior than in the case of the biopore-like cylindrical macropores shown by Schäffer et al. (2008). Although certainly other factors play a role in the deformation of structural pores in natural soils such as inter-particle stabilization by exudates (Zhang et al., 2008) and aggregation (Horn and Dexter, 1989) above discussion underlines the influence of pore space morphology on stress distribution and hence deformation where "biopores" seem to exhibit a more favorable geometry for sustaining stresses compared to shrinkage induced pores. Similar findings were reported by Berli and Or (2006) demonstrating the influence of pore morphology on their deformation where pore closure rates increased with decreasing initial aspect ratio (oblate pores closed faster than spherical pores), and with higher deviatoric stress.

Accessibility of interfaces is a crucial parameter for exchange processes in soils and influenced by structure and pore size distributions (Hartmann et al., 1998). Connected air filled macropores are considered important for oxygen supply sustaining microbial turnover processes. To quantify the "easily" accessible surface area we have triangulated the solid–void interface from the tomograms. Calculated crack surface areas (Fig. 6–12) show a slightly different change with applied stress than porosity values (Fig. 6–10). For the dry condition crack surface area (dashed line in Fig. 6–12) steeply increases when the load increases to 475 N and finally to 950 N, respectively. This is attributed to the development of shear cracks adding surface area which is accessible via the connected main crack volume. Such additional cracks are mainly occurring at the sample rim but also within aggregates (Fig. 6–9). The area between the total macropore surface (solid line in Fig. 6–12) and the crack surface (dashed line in Fig. 6–12) corresponds to the fraction of surface area of disconnected macropore volumes (interior surface within aggregates). We noticed that disconnected macropore surface area was significantly reduced when the load exceeded 475 N. While total surface area only increased slightly some intra-aggregate cracks have joined to the main crack volume increasing the fraction of the connected crack surface area. A pronounced increase in total surface area could be expected due to the occurrence of new shear cracks resulting from loading. However, this was apparently compensated by a reduction of internal surface area associated with a compression of smaller disconnected pore volumes. Analyzing the change in surface area as a function of the removal of successively larger disconnected pore volumes from the surface area calculations we found that for the dry unloaded condition disconnected pore clusters with a size >100 voxels ($\geq$1.3 × 10^{-2} mm³) significantly contributed to the total surface area of the sample (Fig. 6–13). This was not the case after loading the sample with 950 N where the larger intra-aggregate macropores were reduced by compression (dashed line in Fig. 6–13). For the wet condition an overall trend of increasing surface area with increasing load up to 475 N was observed. Like for the dry condition this increase resulted from soil fracturing which generated additional cracks. Finally, when the soil was loaded in wet condition with 950 N the soil structure seemed to collapse, subsequently reducing the surface area of both crack and intra-aggregate macropores (Fig. 6–12).

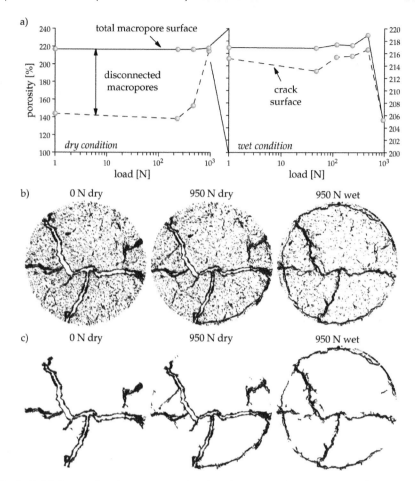

Fig. 6–12. (a) Change of total macropore and crack surface area with mechanical load for the dry condition (left) and wet condition (right). A cross-sectional view of the total macropore surface is shown in (b) and of only the interconnected crack surface area is shown in (c) for different moisture/loading conditions.

As it can be seen already from the tomograms in Fig. 6–9 the deformation process can be described as a combination of volume change (predominantly compression) and shear displacement (fracturing) leading to changes in inter- and intra-aggregate pore space characteristics (connectivity, pore cluster size distribution, pore-void interface surface area). Wetness during loading had a clear influence on deformation by weakening the grain fabric (inter-particle strength). Localized strain analysis by image correlation revealed zones in the sample with different deformation characteristics and magnitudes. We found an overall very heterogeneous deformation behavior at various scales which changed as a function of wetness (Fig. 6–14). For the dry sample equivalent strain ε_{equ} (Eq. [6]) as a

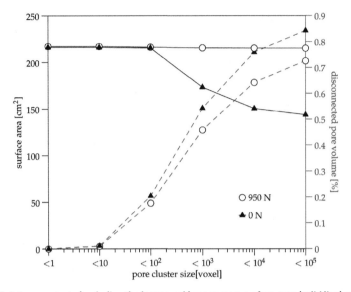

Fig. 6–13. Intra-aggregate (excluding the large crack) macropore surface area (solid line) and disconnected pore volume (dashed line) as a function of pore cluster size before and after loading the sample in dry condition.

scalar quantity reflecting the magnitude of deformation irrespective of the orientation of the principle strain axis showed that deformation was mainly associated with the cracks with about one order of magnitude smaller strains within the aggregates (Fig. 6–14a and 6–14c). However, also within aggregates a patchy pattern of the strain values was observed indicating local differences in soil stiffness/ strength. Wetting the sample to –6 kPa matric potential changed the deformation pattern in a way that now aggregates were more sensitive with respect to mechanical loading with a different strain pattern. This possibly reflects spatial variation of water contents where zones of higher deformation may be associated with higher moisture content (Fig. 6–14b). For the lower loads (≤237 N) under wet condition the deformation within cracks seemed to be less pronounced than under dry condition. This may partly be attributed to the stress history of the sample which has been pre-loaded (loading under dry condition) resulting in a new stress-strain equilibrium due to deformation increasing elasticity of the soil. However, more mechanical energy seems to be dissipated by deformation of the aggregates itself under wet condition (Fig. 6–14b) instead of distributing the stresses toward the cracks, as it was observed for the dry soil condition (Fig. 6–14c). As soon as the loads increased beyond the pre-compression stress deformation for the wet condition was significantly higher in all parts of the sample compared to the dry condition (load of 475 N in Fig. 6–14c and 6–14d). Note that

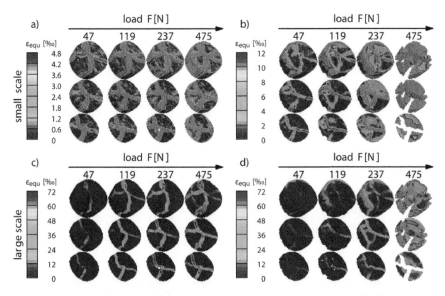

Fig. 6–14. Localized equivalent strain (ε_{equ}) for two different deformation scales. (top) Small scale deformation within aggregates excluding the crack region; (bottom) large scale deformation predominantly occurring in the crack region. Deformation for the dry samples is shown in (a) and (c) and for the wet sample in (b) and (d), respectively.

deformation calculations by three-dimensional digital image correlation was not possible for the crack volume at very high loads for the wet sample indicated by missing cubes in Fig. 6–14 since deformation was too high.

As the sample was subjected to compressive forces we would expect volume reduction and shortening in the direction of the major principle stress which was vertical. Volume changes (dV/V) and axial strain (γ_{zz} representing vertical strain) were affected by the presence of the large crack (Fig. 6–15). Similar to the equivalent strain shown in Fig. 6–14 we also observed heterogeneous distributions of axial strain (γ_{zz}) and volume change (dV/V) in the sample. Moreover, we found localized patterns with simultaneous volume expansion/longitudinal extension (positive values for dV/V and γ_{zz} in Fig. 6–15) and compression/longitudinal shortening (negative values for dV/V and γ_{zz} in Fig. 6–15), respectively. The localized deformation patterns were not completely stable in position but changed with magnitude of load, however, heterogeneity was most pronounced in the vicinity of the crack. At the beginning of the deformation (lower loads) longitudinal extension was dominant in most parts of the sample except in the crack region where also longitudinal shortening occurred. With increasing load more zones were affected by shortening but closely associated with simultaneous expansion leading to steep deformation gradients in the crack vicinity. In the wet condition axial strain was slightly more homogeneous than in the dry condition, neverthe-

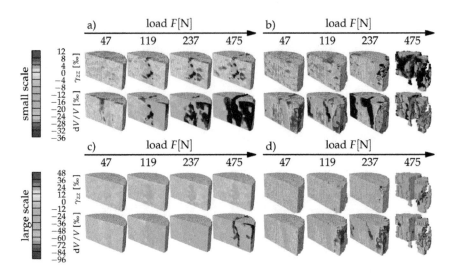

Fig. 6–15. Localized vertical strain (γ_{zz}) and volume strain (dV/V) for two different deformation scales. (top) Small scale deformation within aggregates excluding the crack region; (bottom) large scale deformation predominantly occurring in the crack region. Deformation for the dry samples is shown in (a) and (c) and for the wet sample in (b) and (d), respectively.

less, simultaneous extension and shortening prevailed also here (Fig. 6–15a and 6–15b). At lower loads some parts of the sample showed volume expansion (red colors in Fig. 6–15a) immediately next to the crack where compression was strongest. With increasing load more and more of the sample changed from volume expansion to compression, however, interestingly when the soil was wet some regions with volume expansion remained even at higher loads (Fig. 6–15b and 6–15d). Note that different localized volume changes express also different local changes in porosity and pore size distribution.

The local distribution of the strain tensors shown for a horizontal and vertical cross-section in Fig. 6–16 for a load of 237 N, demonstrates that deformation is not only characterized by the heterogeneous distribution of strain parameters but is in addition anisotropic. Color coding in Fig. 6–16 corresponds to the values for γ_{zz} shown in Fig. 6–15. For mechanical soil deformation in dry condition (Fig. 6–16a) the major principle axes of the strain tensor indicates a preferred vertical orientation in the crack region and a transition toward a preferred horizontal orientation in the aggregates with a sigmoidal strain trajectory similar to what is found in shear zones of deformed rocks (Ramsay and Huber, 1983). Note that although the major principle axes of the strain tensor seems to be associated with a lengthening in the vertical direction within the crack region the colors of γ_{zz} indicate that shortening of the major axis has taken place which is also demonstrated in Fig. 6–15a (negative values for γ_{zz}). This suggests that vertical mechanical loading is associated with sig-

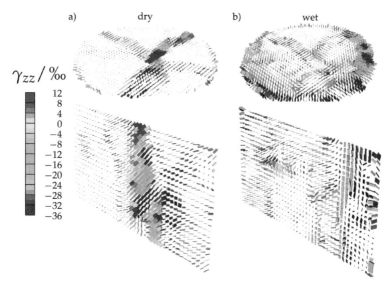

Fig. 6–16. Spatial orientation of the Lagrangrian strain tensor γ represented as cuboids colored according to the tensor entry γ_{zz} for a cross-sectional view (top) and longitudinal-section view (bottom) for the (a) dry and (b) wet condition, respectively.

nificant horizontal deformation (i.e., perpendicular to γ_{zz}) in the crack region (see also ε_{equ} in Fig. 6–14) resulting in a stronger shortening of the two horizontal components of the strain ellipsoid (γ_{xx} and γ_{yy}) compared to the vertical component (γ_{zz}). Shortening of all principle strain components is also confirmed by the volumetric strain which indicated strong compression in the crack region (Fig. 6–15a).

Strain tensors derived for the same magnitude of load but applied under wet condition reveals a different geometry ranging from almost vertically oriented and parallel aligned strain ellipsoids to disordered and rotational strain trajectories (Fig. 6–16b). The different deformation behavior of the sample may partly be attributed to the presence of water but is probably even more affected by the modified architecture of the pore space where, compared to the dry condition, additional shear cracks resulting from previous loading events existed (Fig. 6–9). The deformation process seems to be particular influenced by numerous fractures that are present at the rim of the sample where volume expansion indicates local shear failure (Fig. 6–15d). This demonstrates the impact of soil structure on soil mechanical processes which are highly heterogeneous and anisotropic. The associated changes in pore size distribution and reorientation of pores finally leads to spatially dependent anisotropic transport properties of the pore network (Dörner and Horn, 2009).

Besides soil structure, water had a very clear effect on overall deformation of the sample. This is demonstrated by integrating local strain values calculated from

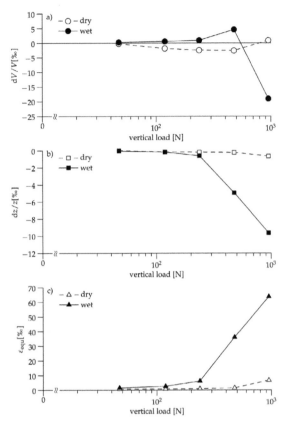

Fig. 6–17. Global deformation indicators for the dry (dashed line) and wet (solid line) condition derived from averaging local strain values: (a) dV/V = volume strain (compression), (b) dz/z = relative vertical displacement, and (c) ε_{equ} = equivalent strain.

digital image correlation results thus providing global mean deformation indicators (Fig. 6–17). Note that integration excludes regions with "bad voxels," which represent problematic areas where convergence of the algorithm was not achieved, being mainly associated with the large crack volume. The change in vertical strain dz/z with load in Fig. 6–17 corresponds well for the wet condition to the external measurement of vertical displacement shown in Fig. 6–8. The strongest change in dz/z was observed close to the precompression stress P_c. Due to the compression of the sample we would have expected a negative volume change dV/V which in contrary was positive probably resulting from the generation of microcracks. This mechanical behavior is clearly an effect of the presence of the large crack providing space for lateral soil movement following shear deformation. In other words especially structured soils could behave internally different than we would assume by only measuring external displacements as we do in oedometer tests. When the load exceeds the precompression stress then the sample seems to fracture indicated by increasingly positive values for dV/V (volume expansion). Only at a very high load of 950 N the structure finally collapses and dV/V becomes negative showing that compression becomes the dominant factor for deformation which is also confirmed by the change in pore size distribution (Fig. 6–11). Equivalent strain (ε_{equ}), however, steeply increases at loads just higher than P_c which is suggested to be mainly the result of closing major cracks by lateral plastic flow and the beginning compression of the intra-aggregate macropores (Fig. 6–11). The latter is not in contradiction to the volume expansion shown by dV/V (Fig. 6–17a) but demonstrates that it is possible to both compress pores in one region and simultaneously generate new pores (micro-cracks) due to ongoing shear deformation in other regions. For the dry condition changes in scalars (dV/V, dz/z, and ε_{equ}) are much less pronounced than for the wet condition. This is partly owing to the fact that loads were not applied homogeneously over the soil surface but concentrated to the outer part of the sample because of the uneven surface left after shrinkage. However, even though stresses must have been very high the overall structure remained remarkably stable. In contrast to the wet condition the sample deforms more compressive at lower loads and only shows volume expansion at a very high load of 950 N which corresponds to the development of new shear cracks (Fig. 6–9).

The analysis of tomographic images following soil deformation by three-dimensional digital image correlation revealed very complex and various spatially distributed deformation mechanisms that cannot be readily deduced from global deformation values as they are usually obtained from soil mechanical tests (Fig. 6–8). The findings presented here confirm results reported by Sun et al. (2004) for silty clay under triaxial loading who documented based on CT-imaging closure and new development of micro-cracks in different deformation stages. They noted that the simultaneous initiation of new micro-cracks and closure of a previously existing macro-crack hint at an anisotropic deformation behavior and that damage evolution in the sample is inhomogeneous. Examining CT images of triaxially compressed fine-grained sand specimens also Alshibli et al. (2006) revealed a very complex deformation pattern with shear bands. Although the initial structure of the sample was less complex than in our study they also found similar to our results volume increases during compression even at axial strain levels as high as 25% suggesting that macroscopic compression not necessarily leads to reduction in porosity at every place in the sample. Complex patterns of strain localization were also observed by Viggiani et al. (2004) concluding that dilating shear bands and cracks were formed and interacted during the course of deformation resulting in abrupt changes in soil density from zones of localization.

Conclusions

Deformation phenomena in soils are influenced by hydraulic and mechanical stresses where the latter mostly acts in combination with hydraulic stresses suggesting that deformation behavior of soils should be treated as a coupled mechano-hydraulic process. Traditional methods like pore size distribution derived from water retention functions and stress-strain relationships obtained from mechanical soil tests represent overall bulk soil properties and behavior but do not reveal the full complexity of soil structure dynamics and deformation processes in soils. X-ray microtomography in combination with image analysis methodologies such as three-dimensional morphological image quantifications and digital image correlation was applied to provide more detailed insight into changes in pore space geometries in soils subject to hydraulic and mechanical stresses. A complex structure was formed initially by shrinkage and water menisci forces which strongly influenced soil deformation by subsequent mechanical loading resulting in localized anisotropic and spatially heterogeneously distributed deformation fields. Consecutive and simultaneous generation and closure of cracks, fissures and intra-aggregate macro-pores determine the change in pore geometry. Mechanical deformation can be described as a combination of compression, volume expansion, shear displacement and plastic flow where both water and structure influences the local deformation processes. Although to this point it is not possible to generalize the mechanical behavior described in this study as it was restricted to one sample at multiple stress conditions, we assume that complex structured soils with cracks will be subject to a composition of the above described deformation mechanisms. This will have to be proved in future studies where not only replicate samples but also the effect of various initial and boundary conditions will have to be systematically investigated to derive a fundamental description of the internal stress-strain relationships including the coupling of mechano-hydraulic processes. Such a more detailed knowledge of deformation patterns related to shrinkage and mechanical loading is needed to develop models linking soil deformation with pore functions to better understand and predict the influence of soil structure dynamics on soil processes.

Acknowledgments

In the scope of the DFG projects INST 212/209-1 FUGG and INST 257/320-1 AOBJ two high performance microfocus computertomography systems, which were deployed in this work, could be procured, which is gratefully acknowledged.

References

Adler, P.M. 1995. Multiphase flow in porous media—Preface. Transp. Porous Media 20:1. doi:10.1007/BF00616922.

Alshibli, K.A., S.N. Batiste, S. Sture, and M. Lankton. 2006. Micro-characterization of shearing in granular material using computed tomography. In: J. Desrues, G. Viggiani, and P. Bésuelle, editors, Advances in x-ray tomography for geomaterials. ISTE, London. p. 17–34.

Baumgartl, T., and B. Köck. 2004. Modeling volume change and mechanical properties with hydraulic models. Soil Sci. Soc. Am. J. 68:57–65.

Baveye, P. 2010. Bacterial growth kinetics and substrate bioavailability in structured soils: Need for a paradigm "shift". In: L.W. De Jonge, P. Moldrup, and A.L. Vendelboe, editors, Proceedings of the 1st International Conference and Exploratory Workshop on Soil Architecture and Physico-Chemical Functions "CESAR," 30 Nov. to 2 Dec., Tjele, Denmark. Faculty of Agricultural Sciences, Aarhus University. p. 426.

Berli, M., and D. Or. 2006. Deformation of pores in viscoplastic soil material. Int. J. Geomech. 6:108–118. doi:10.1061/(ASCE)1532-3641(2006)6:2(108).

Bittelli, M., F. Ventura, G.S. Campbell, R.L. Snyder, F. Gallegati, and P.R. Pisa. 2008. Coupling of heat, water vapor, and liquid water fluxes to compute evaporation in bare soils. J. Hydrol. 362:191–205. doi:10.1016/j.jhydrol.2008.08.014.

Bloomenthal, J. 1988. Polygonization of implicit surfaces. Comput. Aided Geom. Des. 5:341–355. doi:10.1016/0167-8396(88)90013-1.

Bradford, S.A., and S. Torkzaban. 2008. Colloid transport and retention in unsaturated porous media: A review of interface-, collector-, and pore-scale processes and models. Vadose Zone J. 7:667–681. doi:10.2136/vzj2007.0092.

Capowiez, Y., S. Sammartino, and E. Michel. 2011. Using X-ray tomography to quantify earthworm bioturbation non-destructively in repacked soil cores. Geoderma 162:124–131. doi:10.1016/j.geoderma.2011.01.011.

Casagrande, A. 1936. The determination of pre-consolidation load and its practical significance. In: Proceedings of the international conference on soil mechanics and foundation engineering, Harvard University, Cambridge, MA. p. 60–64.

Crostack, H.-A., J. Nellesen, G. Fischer, U. Weber, S. Schmauder, and F. Beckmann. 2008. 3D Analysis of MMC microstructures and deformation by μCT and FE simulations. In: S.R. Stock, editor, Developments in X-ray tomography VI, Vol. 7078. SPIE, San Diego, CA. 70781I p. 1–12.

Darrah, P.R., D.L. Jones, G.J.D. Kirk, and T. Roose. 2006. Modelling the rhizosphere: A review of methods for 'upscaling' to the whole-plant scale. Eur. J. Soil Sci. 57:13–25. doi:10.1111/j.1365-2389.2006.00786.x.

Dexter, A.R. 1991. Amelioration of soil by natural processes. Soil Tillage Res. 20:87–100. doi:10.1016/0167-1987(91)90127-J.

Dexter, A., B. Kroesbergen, and H. Kuipers. 1984. Some mechanical properties of aggregates of top soils from the IJsselmeer polders, II. Remoulded soil aggregates and the effects of wetting and drying cycles. Neth. J. Agric. Sci. 32:215–227.

Dörner, J., and R. Horn. 2009. Direction-dependent behaviour of hydraulic and mechanical properties in structured soils under conventional and conservation tillage. Soil Tillage Res. 102:225–232. doi:10.1016/j.still.2008.07.004.

Feeney, D.S., J.W. Crawford, T. Daniell, P.D. Hallett, N. Nunan, K. Ritz, M. Rivers, and I.M. Young. 2006. Three-dimensional microorganization of the soil-root-microbe system. Microb. Ecol. 52:151–158. doi:10.1007/s00248-006-9062-8.

Gerke, H.H. 2006. Preferential flow description for structured soils. J. Plant Nutr. Soil Sci. 169:382–400. doi:10.1002/jpln.200521955.

Ghezzehei, T.A., and D. Or. 2003. Pore-space dynamics in a soil aggregate bed under a static external load. Soil Sci. Soc. Am. J. 67:12–19. doi:10.2136/sssaj2003.0012.

Hartmann, A., W. Gräsle, and R. Horn. 1998. Cation exchange processes in structured soils at various hydraulic properties. Soil Tillage Res. 47:67–72. doi:10.1016/S0167-1987(98)00074-9.

Hopmans, J.W., and K.L. Bristow. 2002. Current capabilities and future needs of root water and nutrient uptake modeling. Adv. Agron. 77:103–183. doi:10.1016/S0065-2113(02)77014-4.

Horn, R., and A.R. Dexter. 1989. Dynamics of soil aggregation in an irrigated desert loess. Soil Tillage Res. 13:253–266. doi:10.1016/0167-1987(89)90002-0.

Horn, R., H. Domzzal, A. Slowinska-Jurkiewicz, and C. van Ouwerkerk. 1995. Soil compaction processes and their effects on the structure of arable soils and the environment. Soil Tillage Res. 35:23–36. doi:10.1016/0167-1987(95)00479-C.

Horn, R., and S. Peth. 2009. Soil structure formation and management effects on gas emission. Biologia 64:449–453. doi:10.2478/s11756-009-0089-4.

Horn, R., and A.J.M. Smucker. 2005. Structure formation and its consequences for gas and water transport in unsaturated arable and forest soils. Soil Tillage Res. 82:5–14. doi:10.1016/j.still.2005.01.002.

Jégou, D., V. Hallaire, D. Cluzeau, and P. Tréhen. 1999. Characterization of the burrow system of the earthworms Lumbricus terrestris and Aporrectodea giardi using X-ray computed tomography and image analysis. Biol. Fertil. Soils 29:314–318. doi:10.1007/s003740050558.

Lehmann, P., I. Neuweiler, and J. Tölke. 2008. Quantitative links between porous media structures and flow behavior across scales. Adv. Water Resour. 31:1127–1128. doi:10.1016/j.advwatres.2008.08.001.

Lindquist, W.B., S.-M. Lee, W. Oh, A.B. Venkatarangan, H. Shin, and M. Prodanovic. 2005. 3DMA-Rock: A software package for automated analysis of rock pore structure in 3-D computed microtomography images, available at http://www.ams.sunysb.edu/~lindquis/3dma/3dma_rock/3dma_rock.html#Sec_Intro.

Lorensen, W.E., and H.E. Cline. 1987. Marching cubes: A high resolution 3D surface construction algorithm. ACM SIGGRAPH Comput. Graph. 21:163–169. doi:10.1145/37402.37422.

Monga, O., M. Bousso, P. Garnier, and V. Pot. 2008. 3D geometric structures and biological activity: Application to microbial soil organic matter decomposition in pore space. Ecol. Modell. 216:291–302. doi:10.1016/j.ecolmodel.2008.04.015.

Nunan, N., K. Ritz, M. Rivers, D.S. Feeney, and I.M. Young. 2006. Investigating microbial microhabitat structure using X-ray computed tomography. Geoderma 133:398–407. doi:10.1016/j.geoderma.2005.08.004.

Oh, W., and W.B. Lindquist. 1999. Image thresholding by indicator kriging. IEEE Trans. Pattern Anal. Mach. Intell. 21:1–13.

Or, D., and T.A. Ghezzehei. 2002. Modeling post-tillage soil structural dynamics: A review. Soil Tillage Res. 64:41–59. doi:10.1016/S0167-1987(01)00256-2.

Peng, X., R. Horn, S. Peth, and A.J.M. Smucker. 2006. Quantification of soil shrinkage in 2D by digital image processing of soil surface. Soil Tillage Res. 91:173–180. doi:10.1016/j.still.2005.12.012.

Peng, X., R. Horn, and A. Smucker. 2007. Pore shrinkage dependency of inorganic and organic soils on wetting and drying cycles. Soil Sci. Soc. Am. J. 71:1095–1104. doi:10.2136/sssaj2006.0156.

Peth, S. 2010. Applications of microtomography in soils and sediments. In: B. Singh, and M. Gräfe, editors, Synchrotron-based techniques in soils and sediments, Vol. 34. Elsevier, Heidelberg. p. 73–101.

Peth, S., R. Horn, F. Beckmann, T. Donath, J. Fischer, and A.J.M. Smucker. 2008a. Three-dimensional quantification of intra-aggregate pore-space features using synchrotron-radiation-based microtomography. Soil Sci. Soc. Am. J. 72:897–907. doi:10.2136/sssaj2007.0130.

Peth, S., R. Horn, F. Beckmann, T. Donath, and A.J.M. Smucker. 2008b. The interior of soil aggregates investigated by synchrotron-radiation-based microtomography. In: S.R. Stock, editor, Developments in x-ray tomography VI, Vol. 7078. SPIE, San Diego, CA. 70781H p. 1–12.

Peth, S., J. Nellesen, G. Fischer, and R. Horn. 2010. Non-invasive 3D analysis of local soil deformation under mechanical and hydraulic stresses by [mu]CT and digital image correlation. Soil Tillage Res. 111:3–18. doi:10.1016/j.still.2010.02.007.

Pierret, A., C. Doussan, Y. Capowiez, F. Bastardie, and L. Pages. 2007. Root functional architecture: A framework for modeling the interplay between roots and soil. Vadose Zone J. 6:269–281. doi:10.2136/vzj2006.0067.

Raats, P. A.C. 1984. Mechanics of cracking soils. In: J. Bouma and P.A.C. Raats, editors, Proceedings of the ISSS symposium on water and solute movement in heavy clay soils, Aug. 27–31, Int. Soc. of Soil Sci., Wageningen, the Netherlands. p. 23–38.

Raats, P.A.C. 1987. Applications of the theory of mixtures in soil science. Math. Model. 9:849–856. doi:10.1016/0270-0255(87)90003-0.

Ramsay, J.G., and M.I. Huber. 1983. The techniques of modern structural geology. Strain Analysis 1, Academic Press, London, 307 pp.

Richards, L.A. 1931. Capillary conduction of liquids through porous mediums. Physics 1:318–333. doi:10.1063/1.1745010.

Ritchie, J.T., and J.E. Adams. 1974. Field measurement of evaporation from soil shrinkage cracks. Soil Sci. Soc. Am. J. 38:131–134. doi:10.2136/sssaj1974.03615995003800010040x.

Schäffer, B., M. Stauber, T.L. Mueller, R. Müller, and R. Schulin. 2008. Soil and macro-pores under uniaxial compression. I. Mechanical stability of repacked soil and deformation of different types of macro-pores. Geoderma 146:183–191. doi:10.1016/j.geoderma.2008.05.019.

Schrader, S., H. Rogasik, I. Onasch, and D. Jegou. 2007. Assessment of soil structural differentiation around earthworm burrows by means of X-ray computed tomography and scanning electron microscopy. Geoderma 137:378–387. doi:10.1016/j.geoderma.2006.08.030.

Smiles, D.E. 1981. Water relations of cracking clay soil. In: J.W. McGarity, E.H. Hoult, and H.B. So, editors, Proceedings of the properties and untilization of cracking clay soils 5, 24–28 Aug., Univ. of New England, Armidale, New South Wales, Australia. p. 143–149.

Smiles, D.E. 1995. Liquid flow in swelling soils. Soil Sci. Soc. Am. J. 59:313–318. doi:10.2136/sssaj1995.03615995005900020006x.

Smiles, D.E. 2000. Hydrology of swelling soils: A review. Aust. J. Soil Res. 38:501–521. doi:10.1071/SR99098.

Simunek, J., M.T. van Genuchten, and M. Sejna. 2008. Development and applications of the HYDRUS and STANMOD software packages and related codes. Vadose Zone J. 7:587–600. doi:10.2136/vzj2007.0077.

Smucker, A.J.M., and J.W. Hopmans. 2007. Preface: Soil biophysical contributions to hydrological processes in the vadose zone. Vadose Zone J. 6:267–268. doi:10.2136/vzj2007.0057.

Sun, H., J.F. Chen, and X.R. Ge. 2004. Deformation characteristics of silty clay subjected to tri-axial loading, by computerised tomography. Geotechnique 54:307–314. doi:10.1680/geot.2004.54.5.307.

van Genuchten, M.T., F.J. Leij, and S.R. Yates. 1991. The RETC code for quantifiying the hydraulic functions of unsaturated soils. U.S. Salinity Lab., U.S. Dep. of Agric., Riverside, CA, p. 83.

Viggiani, G., N. Lenoir, P. Bésuelle, M. Di Michiel, S. Marello, J. Desrues, and M. Kretzschmer. 2004. X-ray microtomography for studying localized deformation in fine-grained geomaterials under triaxial compression. C. R. Mec. 332:819–826. doi:10.1016/S1631-0721(04)00157-3.

Vogel, H.J., H. Hoffmann, A. Leopold, and K. Roth. 2005. Studies of crack dynamics in clay soil: II. A physically based model for crack formation. Geoderma 125:213–223. doi:10.1016/j.geoderma.2004.07.008.

Weinberger, R. 1999. Initiation and growth of cracks during desiccation of stratified muddy sediments. J. Struct. Geol. 21:379–386. doi:10.1016/S0191-8141(99)00029-2.

Wiermann, C., D. Werner, R. Horn, J. Rostek, and B. Werner. 2000. Stress/strain processes in a structured unsaturated silty loam Luvisol under different tillage treatments in Germany. Soil Tillage Res. 53:117–128. doi:10.1016/S0167-1987(99)00090-2.

Wu, K., N. Nunan, J.W. Crawford, I.M. Young, and K. Ritz. 2004. An efficient Markov Chain Model for the simulation of heterogeneous soil structure. Soil Sci. Soc. Am. J. 68:346–351. doi:10.2136/sssaj2004.0346.

Yoshida, S., and K. Adachi. 2004. Numerical analysis of crack generation in saturated deformable soil under row-planted vegetation. Geoderma 120:63–74. doi:10.1016/j.geoderma.2003.08.009.

Young, I.M., and J.W. Crawford. 2004. Interactions and self-organization in the soil-microbe complex. Science 304:1634–1637. doi:10.1126/science.1097394.

Zhang, B., P.D. Hallett, and G. Zhang. 2008. Increase in the fracture toughness and bond energy of clay by a root exudate. Eur. J. Soil Sci. 59:855–862. doi:10.1111/j.1365-2389.2008.01045.x.

Chapter 7

Gas Permeability in Soils as Related to Soil Structure and Pore System Characteristics

Tjalfe G. Poulsen

Abstract

This chapter presents current knowledge with respect to gas permeability and its relation to soil pore system structure. The presentation is based on a review of the existing literature in the field. Soils are characterized as having (i) a stable and well-connected pore system, (ii) a stable but poorly connected pore system, and (iii) an unstable pore system. The behavior of gas permeability in soils belonging to each of these three types is distinctly different and, therefore, allows for characterization of the pore network structure based on measurements of gas permeability in combination with volumetric air and water contents in the soil. The relation between gas permeability air and water content and pore system characteristics is described and discussed for all three cases based on measured gas permeability data for a variety of soils and other porous materials taken from the existing literature. Mathematical expressions describing the behavior of gas permeability in relation to pore system structure are also presented and their application to the data discussed.

Abbreviations: PO, pore organization index.

T.G. Poulsen, Dep. of Chemistry and Biotechnology, Aalborg Univ., Sohngaardsholmsvej 57, DK-9000 Aalborg, Denmark (tgp@bio.aau.dk).

doi:10.2134/advagricsystmodel3.c7

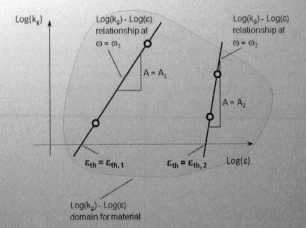

T he ability of soil to conduct air and other gases as a result of pressure gradients is an important property that has implications for a wide range of processes. Remediation of soils contaminated with volatile organics via vapor extraction, formation, transport, degradation and emission of methane at landfills, transport of radon from soil into buildings, and aeration of the soil surface layer and plant root zone are examples of processes that depend on the ability of soil to conduct gas. Gas flow in soils and similar porous media is strongly affected by the structure of the soil. This chapter gives an overview of how the most common types of soil pore system structures affect gas flow through the soil. Models for predicting soil gas flow properties from soil structural characteristics are presented and identification of soil pore system characteristics based on gas flow data is discussed.

Gas Permeability and Soil Structure

Soil ability to conduct gas is generally quantified by its gas permeability, k_g, which has the dimension [L^2], usually μm^2. Conduction of gases in soil in response to pressure gradients has traditionally been described using Darcys law (Freeze, 1994) which for one-dimensional gas flow (neglecting gravitation and assuming laminar gas flow) is given as:

$$\frac{\partial P}{\partial x} = -v\frac{\eta}{k_g} \tag{1}$$

where v is the so-called Darcy (or superficial) velocity [$L\ T^{-1}$], η is gas viscosity [$M\ L^{-1}\ T^{-1}$], P is gas pressure [$M\ L^{-1}\ T^{-2}$], and x is distance in the direction of flow [L]. In cases where the flow cannot be assumed laminar, the relation between v and P is nonlinear and the Forcheimer equation (Gostomski and Liaw, 2001) should be used instead:

$$\frac{\partial P}{\partial x} = -v\frac{\eta}{k_g} - C\rho v^2 \tag{2}$$

where C [L] is a form coefficient that depends on the soil pore system configuration, and ρ is the gas density. Although both k_g and C are important in determination of gas flow in porous media, this chapter will focus solely on k_g, and thus, on gas flow under mainly laminar gas flow conditions. The gas conductivity (equivalent to hydraulic conductivity) can be determined as

$$K_g = \frac{g \rho k_g}{\eta} \qquad\qquad [3]$$

where K_g is the gas conductivity [L T^{-1}] and g [L T^{-2}] is the gravitational acceleration. It is noted here that there is at present not full consensus about the use of the two terms: Gas permeability and gas conductivity. Some researchers sometimes also refer to K_g as gas permeability. The majority of studies however are referring to the two parameters as K_g (conductivity) and k_g (permeability) and this notation is also adopted here.

In the laboratory gas permeability in a given soil is traditionally determined by measuring the pressure drop as a function of gas flow through a cylindrical sample with known length of the soil followed by fitting either Eq. [1] or [2] (or other equivalent expressions) to the measured data to obtain k_g. In situ determination of k_g in the field is usually conducted by blowing air into the soil via a probe (which could be a cylinder or perforated tube) inserted into the soil at selected flow rates and measuring the corresponding pressure drop. Gas permeability is then determined by fitting Eq. [1] or [2] (modified to represent the gas flow field in the soil) to the measured data. More details about how to measure gas permeability is given in Iversen et al. (2001).

Soil gas permeability is strongly dependent on soil structure, especially the continuity of the active (or functional) gas-filled pore system, and is very sensitive to even small changes in soil structure and pore connectivity. Because it is also very simple and fast to measure, gas permeability has been extensively proposed and used as an index of soil structural characteristics such as pore continuity and active pore diameter, by several researchers including Buehrer (1932), Kirkham et al. (1958), Grover (1955), Green and Fordham (1975), Ball (1981), Ball et al. (1988), Fish and Koppi (1994), Schjønning et al. (2002) and Schjønning et al. (2005). Gas permeability has been used for characterizing pore system continuity in both dry and unsaturated soils and also in frozen soils where the use of water infiltration is not feasible (Saxton et al., 1993).

Ball (1981) derived a theory for relating gas permeability, pore system tortousity, and the average diameter of the gas conducting pores in soil.

$$r = \sqrt{\frac{8k_g}{D/D_0}} \qquad\qquad [4]$$

where r [L] is the average (or equivalent) pore diameter of the connected, gas conducting pores, D [L^2 T^{-1}] is the gas diffusion coefficient in the soil gas-filled pore system and D_0 [L^2 T^{-1}] is the gas diffusion coefficient in free air. The relative gas diffusivity, D/D_0, is taken as a measure of soil pore system tortousity. In addition to gas permeability, the equivalent pore diameter has also been used as an indicator of soil pore connectivity.

Groenevelt et al. (1984) introduced the ratio between gas permeability and gas-filled porosity, ε [$L^3 L^{-3}$], as an index for how gas-filled pores were organized and connected within the pore system. Two approaches for calculating this pore organization index (PO) were proposed:

$$PO = \frac{k_g}{\varepsilon} \qquad [5]$$

$$PO = \frac{k_g}{\varepsilon^2} \qquad [6]$$

where k_g and ε are corresponding values measured at the same water content. High values of PO indicate that the soil has a system of well-connected, highly conductive (large) gas-filled pores while total gas filled porosity is low. Low values indicate either that the soil pores are poorly connected or that the soil pores may be well connected but small, both resulting in restricted gas flow.

The slope of the ε–k_g relationship, $dk_g/d\varepsilon$, was proposed as another measure of pore connectivity and pore organization as a function of ε by Poulsen and Moldrup (2006). High values of $dk_g/d\varepsilon$ indicate that the pore system is becoming increasingly connected for increasing values of ε, whereas values of $dk_g/d\varepsilon$, near zero indicates that pore system connectivity is constant and independent of ε.

Several terms have been used throughout the literature when describing the level of structure in a soil (Schjønning, 1986; Moldrup et al., 2003; Poulsen et al., 2008): Highly structured, poorly structured, unstructured and well structured just to mention a few. In general these terms do not give much information as to the nature of the soil structure for instance in terms of pore system connectivity and pore size distribution. Two "highly structured" soils may very well have very different pore connectivity simply because the nature of the structure in the two soils is different. Thus, it is better to define soil structure in terms of pore system characteristics. Soil pore systems have often been described by their pore size distribution and their ability to retain or transmit fluids (such as gas or water). As a key function of the soil pore system is to conduct or retain fluids, soil pore system characteristics will be discussed in terms of their ability to conduct gases or liquids (also sometimes termed the pore system connectivity) and retain liquid rather than in terms of pore size distribution and specific quantity of pores above or below a given diameter.

It is not possible to directly assess soil structure based only on the absolute value of gas permeability measured under certain conditions. The pores in a sand soil and a clay soil with identical water contents can be equally well connected and still the gas permeability of the clay can be orders of magnitude smaller than for the sand simply because the gas-filled pores are smaller. Thus, conclusions based only on one value of gas permeability should therefore be made only for soils with the same texture. Gas permeability in homogeneous soils of different textures under dry conditions could be used to assess the level of structure (pore connectivity, aggregation, or compaction) in soils of the same texture. Table 7–1 lists some typical ranges of gas permeability in homogeneous soils of different textures under dry conditions. These values are averages based on observations presented in the existing literature in the field.

The values in Table 7–1 can be compared to measurements of gas permeability for structured soils and used to evaluate if the soil has better pore connectivity pores than the corresponding homogeneous soil. It does not, however, give any information about how the pore system is connected compared to the homogeneous soil. More information about the pore system in a given soil can be gained from the relationship between gas permeability and gas-filled porosity (ϵ) under variably saturated conditions. The k_a–ϵ relationship has been used by several researchers for characterizing soil pore system connectivity, for example, Ball et al. (1988) Poulsen et al. (2001), Schjønning et al. (2002), Moldrup et al. (2001, 2003), Schjønning et al. (2005), Tuli et al. (2005), Kawamoto et al. (2006), Poulsen and Moldrup (2006), Poulsen et al. (2007), Resurreccion et al. (2007), and Poulsen et al. (2008).

Based on these and other studies of the k_a–ϵ relation in differently textured and structured soils, it is possible to divide soils into three categories with respect to structure.

1. Soils with stable well connected pore systems where the pore system gradually becomes available for gas flow as water is drained from the soil. These soils typically have a uni-modal pore size distribution where all pore sizes in the soil are well represented. In such soils all pores conduct air in proportion to their diameter.

Table 7–1. Approximate ranges for gas permeability in homogeneous soils of different texture based on data from the existing literature.

Soil type	k_g
	μm^2
Clay	<1
Silt	1–10
Fine sand	10–100
Coarse sand	100–1000
Gravel	>1000

2.   Soils with stable but poorly connected pore systems where large parts of the gas-filled pore system becomes connected only below a specific water content or do not become connected at all regardless of water content. These soils are often characterized as aggregated and often have multi-modal (often bi-modal) pore size distributions where certain pore sizes are not well represented. In these soils pores do not necessarily conduct gas in proportion to their diameter due to lack of connections between pores.

3.   Soils with an unstable pore system with variable pore connectivity. These are usually soils that exhibit swelling, shrinking or formation of weaker or stronger aggregates in response to changes in water content. It can also be soils that have been tilled or otherwise disturbed and are changing toward a more natural and stable structure.

These three categories of soil structure and the associated relationships between gas permeability and soil gas-filled porosity are discussed in more detail in the following sections. In addition to these structural categories, soils are also often observed to be anisotropic with respect to gas permeability. Anisotropy is often encountered in layered soils where gas permeability in the horizontal directions can be orders of magnitude larger than in the vertical direction much like what is observed for hydraulic conductivity in layered soils. Soils that are anisotropic with respect to hydraulic conductivity are very likely also anisotropic with respect to gas permeability and gas permeability can therefore with advantage be used to assess soil anisotropy rather than hydraulic conductivity as measurements of gas permeability are much more rapid and simple to carry out (Iversen et al., 2001). In fact gas permeability in dry soil or soil drained to field capacity could be used to assess the saturated hydraulic conductivity. This has been investigated by Schjønning (1986) and Loll et al. (1999) who proposed expressions for calculating saturated hydraulic conductivity from measurements of gas permeability. As the impact of anisotropy on gas permeability is very similar to that of anisotropy on hydraulic conductivity which in turn is well described in the literature, anisotropy will not be discussed further in this chapter.

A significant part of the literature in the field have focused on development of expressions for predicting the relationship between k_g and various soil pore system properties including ε, (a more detailed review of these are presented in the following) based on simple multi-parameter regression procedures using least squares. Applicability of the expressions has also generally been assessed based on sums of squared deviations between measured and predicted values. More in-depth statistical investigations such as for instance determination of confidence intervals for empirical constants or comparison of measured values from different locations are generally very few. The main reason is likely that data often are too limited to perform such investigations based on common statistical methods and also that available data do not fulfill the requirements for using common statistical methods. In general common widely known analytical statistical methods are based on the assumptions that data comes from a normal distribution and that they are independent (except when paired data are measured). These requirements, however, are often not met by field data, meaning that distribution free (non-parametric) statistical methods need to be applied instead. These methods are often not well known and, thus, statistical investigations are few. A review of some useful methods such as bootstrapping for estimation of confidence intervals or ranking based tests for comparison of means can be found in Corder and Forman (2009).

Gas Permeability in Soils with a Stable and Well Connected Pore System

In soils where the structure is stable, and thus, the pore system and pore connectivity does not change significantly over time, changes in gas permeability are primarily associated with changes in the soil water content. If the soil has a well connected pore system, both the soil water retention curve and the relationship between gas permeability and gas-filled porosity will be smooth without any abrupt changes in curve slope. This does not mean that gas permeability cannot vary over several orders of magnitude as a function of gas-filled porosity in the same soil, or as a function of texture across different soils. Stable structure and well connected pore systems are often found in granular (sandy) soils that are not disturbed for instance via agricultural activities such as tilling and sowing. These soils also often have a uni-modal pore size distribution.

Gas Permeability–Gas-filled Porosity Relations in Soils with a Stable and Well Connected Pore System

Measurements of gas permeability in soils with stable and well connected pore systems indicate that the relationship between gas permeability and gas-filled porosity often is linear in a $\log(\varepsilon)$–$\log(k_g)$ plot such that

$$\log(k_g) = A\log(\varepsilon) + B \qquad [7]$$

where A and B are constants. This observation has been done by several researchers, for example, Groenevelt et al. (1984), Ahuja et al. (1984), Ball et al. (1988), Blackwell et al. (1990), Moldrup et al. (1998), Poulsen et al. (1998, 2001) Schjønning et al. (2002), Tuli et al. (2005), Kawamoto et al. (2006), Poulsen et al. (2007). These measurements also indicate that there is a critical value of gas-filled porosity (ε_{cr}) below which the gas-filled pore system is disconnected and the gas permeability essentially is zero. The value of ε_{cr} is, thus, the gas filled porosity at which the gas-filled pore system begins to connect such that gas can begin to flow through the soil.

Although the pore-size distribution with a high degree of accuracy can be determined from the soil water retention curve—pF = log[−soil water potential, ψ (cm)] as a function of the volumetric soil water content θ ($L^3 L^{-3}$)—it does not necessarily mean that the soil water retention curve and the $\log(\varepsilon)$–$\log(k_g)$ relationship share the same degree of linearity again supporting that pore size distribution is less important than pore connectivity. This is illustrated in Fig. 7–1 for two intact soil samples collected at 5 and 75 cm below the soil surface at an experimental field at the Hiroshima University campus, Japan. The values of A and B are 5.3 and 4.6 for the 5-cm depth sample and 6.8 and 6.8 for the 75-cm depth sample, respectively. It is clear that even though the $\log(\varepsilon)$–$\log(k_g)$ relationships for both samples are very linear, only the soil water retention curve [θ–$\log(-\psi)$ relationship] for the 75-cm depth sample is also linear. Thus, the $\log(\varepsilon)$–$\log(k_g)$ can in general not be directly predicted from the shape of the soil water retention curve.

In soils where the ε–k_g relationship follows Eq. [7] (soils with well connected pore systems), the pore organization index, PO, increases monotonically with a progressively increasing slope as a function of ε and this is also the case for $dk_g/d\varepsilon$ as a function of ε. This means that for these soils the gas-filled pore system get progressively better connected as ε increases. Figure 7–2 shows PO and $dk_g/d\varepsilon$ as a function of ε for the soils in Fig. 7–1. In both cases, values based on measurements exhibit progressively increasing relationships and closely match the predicted values.

Modeling Gas Permeability–Gas-Filled Porosity Relations in Soils with a Stable and Well Connected Pore System

Values of B in Eq. [7] have been observed to vary over several orders of magnitude while A typically varies between 1 and 10. Several attempts to develop relations

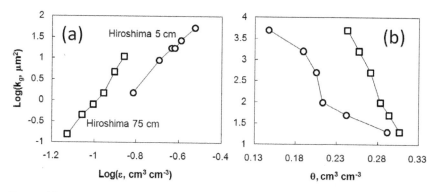

Fig. 7–1. (a) Gas permeability–gas-filled porosity relationships, and (b) soil water retention curves for two intact soil samples collected at an experimental field at the Hiroshima University campus, Japan.

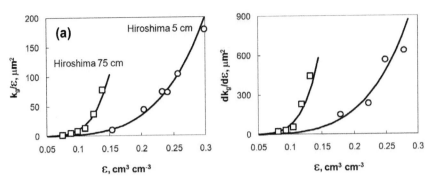

Fig. 7–2. (a) Measured and calculated (based on Eq. [7]) values pore organization index, PO, (as given by Eq. [5]) and (b) $dk_g/d\varepsilon$ as a function of ε for two intact soil samples collected at an experimental field at the Hiroshima University campus, Japan.

between the constants A, B and soil physical properties have been performed to produce a closed-form expression by which the entire k_g–ε relationship for a given soil could be predicted based on simple and easy to measure soil properties.

Because gas permeability is strongly related to gas-filled porosity, it is also related to the water retention properties of the soil. Soils that loose most of their water at low pF values (such as sandy soils) in general have higher gas permeability than soils where higher pF is required to mobilize the water. Moldrup et al. (1998) therefore suggested that A is related to the slope of the soil water retention curve and proposed to use the Campbell (1974) retention parameter, b, as basis for development of an expression for predicting A. Kawamoto et al. (2006) tested several different expressions for A as related to b using gas permeability measurements for several soils and found that

$$A = 1 + \frac{3}{b} \qquad\qquad [8]$$

gave the best predictions of gas permeability as a function of gas-filled porosity. This exponent is equal to the exponent proposed by Alexander and Skaggs (1986) for predicting unsaturated hydraulic conductivity from volumetric water content. As b typically varies between 1 and 20 for most soils (Clapp and Hornberger, 1978) it follows that A in most cases varies between 1 and 4 with the higher values corresponding to coarse (sandy) soils and the lower values to fine (silty and clayey) soils. Higher values of A may be observed in cracked soils.

The gas-filled porosity corresponding to a soil water potential of –100 cm H_2O (pF 2) has been used as an indicator of soil macro-porosity. This also corresponds approximately to the conditions in the soil when it is wetted to field capacity. At this soil water potential pores with diameters at or above 30 μm are gas filled, thus, pores with diameters larger than this value have been regarded as being macro-pores (Schjønning, 1986). Large pores generally conduct most of the gas flowing through a given soil. Based on this, Moldrup et al. (1998) suggested that the constant, B be predicted as follows:

$$B = \frac{k_{g100}}{\varepsilon_{100}^{A}} \qquad [9]$$

where k_{g100} and ε_{100} are corresponding values of k_g and ε measured at a soil water potential of -100 cm H_2O. The point (k_{g100}, ε_{100}) therefore represents a reference point on the $k_g - \varepsilon$ curve. To predict the $k_g - \varepsilon$ relationship using Eq. [7–9] requires knowledge about the soil water retention curve and one measurement of k_g and ε at pF 2. For $A = 1$ or 2, B is equal to one of the two pore organization indices in Eq. [5] and [6] and for soils with a given texture, the constant B could therefore, be regarded as a measure of pore organization and pore system connectivity.

Poulsen et al. (2001) measured surface soil gas permeability in sandy and loamy soil both under laboratory conditions at pF 2 and in the field 24 h after rainfall and found that these values corresponded well. To obtain corresponding values of k_{g100} and ε_{100} in the upper soil under field conditions for sandy and loamy soils, measurements should therefore be taken at least 24 h after rainfall as the soil will have approached field capacity. For deeper soils the time span should be longer. In all cases a minimum depth to the water table of 100 cm is required to allow for drainage.

A number of semi-empirical expressions for predicting k_{g100} have been proposed. Poulsen et al. (1998) investigated the relationship between k_{g100} and ε_{100} for a set of sandy soils and developed an expression based on measurements of gas diffusivity and gas permeability at pF 2 in these soils, combined with the permeability theory of Ball (1981). This expression is given as:

$$k_{g100} = \frac{1300\varepsilon_{100}^{4.45}}{\phi^{3.45}} \qquad [10]$$

where ϕ [$L^3 L^{-3}$] is the soil total porosity and k_{g100} is given in μm^2. Kawamoto et al. (2006) proposed a somewhat simpler model for predicting k_{g100} based only on ε_{100} using measurements from several sandy and loamy soils with clay contents up to 25%. This model is given as

$$k_{g100} = 890\varepsilon_{100}^{2.5} \qquad [11]$$

This expression also predicts k_{g100} in μm^2 and is valid for ε_{100} ranging between 0.1 and 0.4 cm³ cm⁻³. Poulsen et al. (2007) also investigated the relationship between

k_{g100} and ε_{100} for a set of sandy and loamy field soils with clay contents up to 20% and found that for these soils k_{g100} (in μm^2) could be predicted from

$$k_{g100} = 2.5 + 15,830\varepsilon_{100}^{5.2} \qquad [12]$$

Predictions of $\log(k_{g100})$ as a function of ε_{100} for a typical sandy soil with a total porosity of 0.45 cm^3 cm^{-3} by each of the three models (Eq. [10–12] are shown in Fig. 7–3. In general the three models are in fair agreement (within 1 order of magnitude) between ε_{100} = 0.1–0.4 cm^3 cm^{-3} although Eq. [11] predicts the highest values of k_{g100} at low ε_{100} and vice versa at high values of ε_{100}. Given the typical variability in k_g in most natural soils, however, the choice of model for estimating k_{g100} is relatively unimportant as the uncertainty associated with the choice of model generally will be less than the natural variability in k_{g100} in a given soil.

Using Eq. [7–9] in combination with either Eq. [10], [11], or [12] together with one measurement of ε_{100}, enables prediction of the entire k_g–ε relationship for soils with a stable and well connected pore system if the soil water retention curve is known. When using a measured value of k_{g100} together with the soil water retention curve (instead of Eq. [10], [11], or [12]), however, the uncertainty in the predictions of k_g is about 35% of the uncertainty if only the soil water retention curve is used in the prediction (Kawamoto et al., 2006).

Equations [8] and [9] suggest that for a given soil there exist a dependency of B on A and of both A and B on the shape of the soil water retention curve. Poulsen et al. (2007) investigated the relation between A and B for sandy and loamy soils collected from the surface and down to a depth of 7 m and found a linear relation between A and B (Fig. 7–4). Figure 7–4 also shows that the relationship is better if only the subsurface soils collected at depths below 40 cm (Fig. 7–4b) are considered compared to all soils including the surface soils (Fig. 7–4a). This is probably because some of the surface soils have an unstable and possibly less

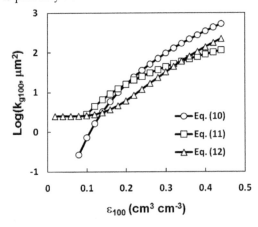

Fig. 7–3. Predictions of k_{g100} from ε_{100} using Eq. [10], [11], or [12] for a typical sandy soil with a total porosity of 0.45 cm^3 cm^{-3}.

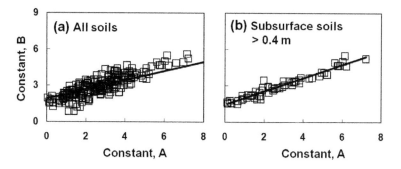

Fig. 7–4. Relationship between the constants *A* and *B* in Eq. [7] for a set of sandy and loamy soils collected at the 0- to 7-m depth. Modified from Poulsen et al. (2007).

well connected soil pore system for instance caused by recent tilling or extreme drying. These soils will likely alter their pore structure in response to the wetting necessary to measure k_g as a function of ε and may therefore exhibit somewhat nonlinear $\log(k_g)$–$\log(\varepsilon)$ relations resulting in the increased scatter observed in Fig. 7–4a. The subsurface soils with a more stable and well connected pore system, however, tend to exhibit a nice linear relationship between *A* and *B*. If the slope of the *A*–*B* relationship is α, and the intercept is β, the relationship between *A* and *B* can be written as

$$B = \alpha A + \beta \qquad [13]$$

This means that Eq. [7] can be written as

$$\log(k_g) - \beta = A \left[\log(\varepsilon) + \alpha\right] \qquad [14]$$

For the soils in Fig. 7–4a, values of the constants α and β were 0.55 and 1.45, respectively, for k_{g100} and ε_{100} in μm^2 and $cm^3\ cm^{-3}$, respectively. The linear relation (Eq. [13]) between *A* and *B* means that all $\log(k_g)$–$\log(\varepsilon)$ relationships in theory will pass through the point $(\varepsilon, k_g) = (-\alpha, \beta)$. For the data in Fig. 7–4 this corresponds to $(\varepsilon, k_g) = (0.28\ cm^3\ cm^{-3}, 28\ \mu m^2)$. The $\log(k_g)$–$\log(\varepsilon)$ relationships for nine randomly selected soils from Fig. 7–4b, are shown in Fig. 7–5 together with the best fit straight lines.

It is seen that the $\log(k_g)$–$\log(\varepsilon)$ relationships with good approximation passes through a common intersection point. A similar observation indicating a common point of intersection for the $\log(k_g)$–$\log(\varepsilon)$ relationships was done for yard waste compost with an organic matter content of 14%, by Poulsen and Moldrup (2007). In this case the common intersection point $(-\alpha, \beta)$ was located at $\log(\varepsilon, cm^3\ cm^{-3})$ $= -0.19$ and $\log(k_g, \mu m^2) = 2.1$ corresponding to $(\varepsilon, k_g) = (0.65\ cm^3\ cm^{-3}, 125\ \mu m^2)$ both of which is higher than for the sandy soils in Fig. 7–5. As the compost also had

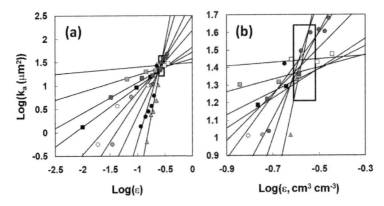

Fig. 7–5. The log(k_g)–log(ε) relationships for nine randomly selected sandy sub-soils from Fig. 7–4b. (a) View of entire log(k_g)–log(ε) range and (b) close-up of the area near the average point of intersection log(ε), log(k_g) = –0.55 cm³ cm⁻³, 1.45 μm². The rectangle indicates the region through which 90% of the relationships for the deep soils (64 samples from Fig. 7–4b) pass. Modified from Poulsen et al. (2007).

higher gas permeability and gas-filled porosity compared to the sandy soil the results indicate that the location of the common point of intersection possibly is proportional to the ranges of k_g and ε for the material in question.

The implications of the results in Fig. 7–4 and 7–5 are that predictions of the entire ε–k_g relationship for a given soil could be made if the location of the common point of intersection (–α, β) and one reference measurement of corresponding ε, k_g values for the soil is known. It remains to be seen, however, if the presence of a common point of interception is valid for all soil types. The slope of the linear log(ε)–log(k_g) relationship (A) is then estimated based on the common point of intersection (–α, β) and the reference measurement point (ε^*, k_g^*) assuming that the relationship intercepts both these points. The slope A is estimated as follows:

$$A = \frac{\log\left(k_g^*\right) - \beta}{\log\left(\varepsilon^*\right) + \alpha} \qquad [15]$$

Combining Eq. [14] and [15] yields

$$\log\left(k_g\right) = \left[\frac{\log\left(k_g^*\right) - \beta}{\log\left(\varepsilon^*\right) + \alpha}\right]\left[\log\left(\varepsilon\right) + \alpha\right] + \beta \qquad [16]$$

The advantage of Eq. [16] is that once the common point of interception (–α, β) has been established for a given soil type, future predictions of the ε–k_g relationship for soils of the same type can be made based on a single reference (ε^*, k_g^*) measurement. To achieve optimal estimates of the relationship it is important that the distance between the points $(-\alpha, \beta)$ and $[\log(\varepsilon^*), \log(k_g^*)]$ is sufficiently large as the uncertainty in the estimates of A (and B) increases with decreasing distance between the points.

Poulsen et al. (2007) suggested that for sandy soils $(\varepsilon_{100}, k_{g100})$ could be used as reference point (ε^*, k_g^*) bearing in mind that the larger the distance between the two points the more accurate the predictions will be in general. Equation [16] could be combined with either Eq. [10], [11] or [12] to yield an expression for predicting the entire $\log(\varepsilon)$–$\log(k_g)$ relationship based only on a single measurement of ε_{100} for the soil in question provided the values of α and β are known. This concept was tested for a set of sandy soils (Poulsen et al. (2007) using $(\alpha, \beta) = (0.19, 2.1)$ in combination with Eq. [12] and [16] and measured values of ε_{100}. Deviations between measured and predicted values of k_g were in general within 1 order of magnitude. Using Eq. [16] in combination with measured values of both ε_{100} and k_{g100} (instead of Eq. [12]) reduced deviations to yield an accuracy of about one-half order of magnitude. Considering the large variability of gas permeability in natural soils this accuracy is acceptable for most purposes.

Gas Permeability in Soils with a Stable but Poorly Connected Pore System

Poorly connected pore systems in soils can in most cases be categorized into two different types: (i) soils with few but well connected, highly conducting (usually large) pores surrounded by regions of comparatively small or poorly connected pores and (ii) soils with an extensive network of large pores that are connected via relatively narrow pore throats (restricting gas or water flow) surrounded by regions of small or poorly connected pores. The layout of the pore systems need not be very different, it is the presence of the narrow pore throats in one system and not in the other that makes the main difference in pore system connectivity and the shape of the $\log(\varepsilon)$–$\log(k_g)$ relationship. Of course not all soils with poorly connected pore systems will fall fully into either one or the other of these categories but may instead be somewhere between them and will therefore have mixed characteristics with respect to k_g–ε relations.

Gas Permeability–Gas-Filled Porosity Relations in Soils with a Limited Network of Well Connected Pores Surrounded by Regions of Small or Poorly Connected Pores

A limited network of well connected and highly conductive (large) pores surrounded by regions with comparatively very small or poorly connected pores are often observed in strongly aggregated (including cracked) soils. It is also often so that the well connected porosity (the total volume of the highly conducting pores) constitutes only a relatively small fraction of the total porosity. These soils will in many cases have bimodal pore size distributions with many small and large pores and few pores of intermediate size. Alternatively the soils may have a well connected system of pores surrounded by regions of pores that are not necessarily much smaller but instead are relatively poorly connected.

In these soils, instead of following a power function such as seen in Fig. 7–1 and Eq. [7], k_g will increase at a more moderate rate as function of ε, when the well connected fraction of the pore system is gas filled. The reason is that per volume of gas-filled pore space, the small well connected part of the pore system can transmit much more gas than the surrounding regions with smaller or poorly connected pores. This means that as soon as the critical value of gas-filled porosity ε_{cr} has been reached, k_g will increase very rapidly with increasing ε, possibly by several orders of magnitude until the entire well-connected pore system has become gas filled. When this has happened k_g will increase at a more moderate rate for increasing ε (k_g may even be constant in this region).

This is illustrated in Fig. 7–6 for two intact samples taken from the same location (Hiroshima) as the data in Fig. 7–1. For both samples in Fig. 7–6, the well-connected pore network is gas filled when pF is larger than 1.7 and k_g increases rather moderately with ε above this point. The data indicate that the well-connected pore network in the Hiroshima soils consists of pores with diameters larger than about 60 µm (the equivalent pore diameter corresponding to pF = 1.7) The well-connected porosity in the samples in Fig. 7–6 is 39 and 65% of total porosity for the 25 and 75 cm samples, respectively, and is, thus, relatively large.

Comparing the retention curves (Fig. 7–6b) with the gas permeability curves (Fig. 7–6a) reveals that the sample taken at 75 cm depth does not have a bimodal pore size distribution and, the pore network in this sample is therefore of the type with a limited but well connected network of pores surrounded by regions of poorly connected pores that are not necessarily much smaller. This is also

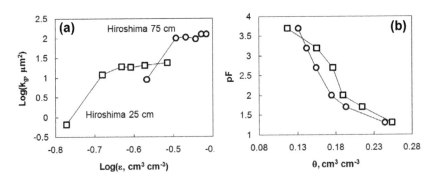

Fig. 7–6. (a) Gas permeability–gas-filled porosity relationships and (b) soil water retention curves, for two intact soil samples collected at an experimental field at the Hiroshima University campus, Japan.

consistent with a well-connected porosity constituting 65% of the total porosity. The 25 cm sample exhibit a tendency for bi-modal pore size distribution with relatively few pores draining between pF 2 and pF 2.7 (Fig. 7–6b) meaning that this sample has relatively few pores with diameters between 6 and 30 μm. This is also the case for the 5 cm sample in Fig. 7–1 even though the pore system in this sample is much better connected even for the smaller pores.

Although the gas permeability of individual aggregates in an aggregated soil is usually lower than the gas permeability of the soil matrix as a whole, gas permeability of individual aggregates can still vary considerably. McKenzie and Dexter (1996) investigated aggregate gas permeability in a sandy loam soil and found that k_g of the individual aggregate varied between 0.1 and 100 μm² depending on water content. This is the same as if the soil was homogeneously packed (Table 7–1). In general there exists no good general relationship between soil overall gas permeability and the gas permeability of individual aggregates in the same soil.

Gas Permeability–Gas-Filled Porosity Relations in Soils with a Network of Larger Pores Connected via Smaller Pore Throats, Surrounded by Regions of Small or Poorly Connected Pores

In soils where part of the system of larger pores is constricted by smaller pore throats, the system will only be partially connected if ε is larger than ε_{cr} but smaller than the value of ε at which the narrow pore throats are drained (or filled) and become available for gas flow. Often the pore throats are of similar size and will therefore drain (or fill) within a relatively narrow range of ε and in this range k_g will increase rapidly with increasing ε as the larger pores become connected. When they are fully connected k_g will tend to vary with ε at a more moderate rate in response to drainage (or filling) of the smaller or poorly connected pores. This is illustrated for two samples from the Hiroshima soil in Fig. 7–7.

Taking the sample collected at 5 cm depth as an example it is seen that for ε ranging between 0.25 and 0.35 cm³ cm⁻³ (pF 1.3–2.0), k_g increases rapidly with ε indicating that the pore system is becoming increasingly connected in this region. For ε = 0.35–0.42 cm³ cm⁻³ (pF = 2.0–3.2), the ε–k_g relationship has a plateau where k_g stays constant indicating that the connectivity of the pore system is not increasing (no important pore connections are being made) while above ε = 0.42 cm³ cm⁻³, k_g increases again indicating that the pore system once again is becoming increasingly connected. The slight decrease in k_g at (pF 2.0–2.7) could indicate a very slightly unstable pore system (about this later), however, in comparison with the uncertainty associated with measuring k_g in undisturbed soil, it is too minor for any firm conclusions.

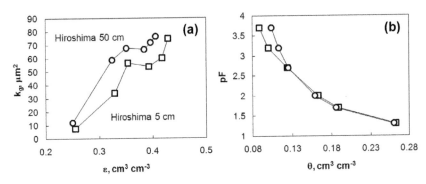

Fig. 7–7. (a) Gas permeability–gas-filled porosity relationships and (b) soil water retention curves for two intact soil samples collected at an experimental field at the Hiroshima University campus, Japan. Note that the *y* axis in Fig. 7–7a is not logarithmic.

The data in Fig. 7–7a indicate that the pores controlling connectivity are of two size ranges, 30–150 μm and <6 μm. This likely means that the sample has a pore network consisting of large pores with diameters ranging between 30 and 150 μm. Many of these pores are directly connected but a significant fraction of the pores are connected via pore throats that are less than 6 μm in diameter (corresponding to pF > 3.2). For both soil samples in Fig. 7–7, the large pores accounts for approximately 65% of the total porosity.

The soil samples in Fig. 7–7 have a system of larger pores that is connected via small pore throats with sizes belonging to one size interval but in theory the large pores could be connected by smaller pores with sizes belonging to more than one size interval. In such cases the soils will exhibit ε–k_g relationships with multiple plateaus as illustrated in Fig. 7–8. It could be argued that the ε–k_g relationships shown in Fig. 7–1, 7–6, and 7–7 are special cases of the more general ε–k_g relationship illustrated in Fig. 7–8.

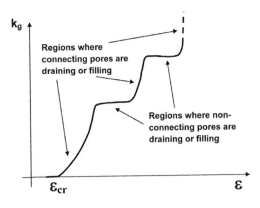

Fig. 7–8. Generalized ε–k_g relationship illustrated for a pore system consisting of macro-pores partly connected via smaller pores with diameters belonging to two different pore diameter intervals.

The ε–k_g relationships for soils with well connected pore systems (such as those in Fig. 7–1) will correspond to the part of the theoretical ε–k_g curve from ε_{cr} to the first plateau in the curve. The ε–k_g relationship for soils with a limited but well connected system of large pores surrounded by regions of small or poorly connected pores (such as those in Fig. 7–6) will correspond to the part of the theoretical ε–k_g curve from ε_{cr} to and including the first plateau in the curve. Finally soils with a network of larger pores, poorly connected via small pore throats surrounded by regions of small or poorly connected pores with the connecting pore throats having diameters belonging to one size interval will correspond to the part of the theoretical ε–k_g curve from ε_{cr} to and including the second region of rapid increase in k_g (including the first plateau).

Modeling Gas Permeability–Gas-Filled Porosity Relations in Soils with a Stable but Poorly Connected Pore System

Poulsen and Moldrup (2006) proposed a concept for modeling the ε–k_g relationship in soils with with a system of large pores connected via pore throats of smaller size and surrounded by regions of small or poorly connected pores, corresponding to the ε–k_g relationships presented in Fig. 7–6. This model was developed based on data for a set of volcanic soils from Japan using theory presented by Brutsaert (1966) and van Genuchten (1980). The expression proposed by Poulsen and Moldrup (2006) is

$$k_g = \sum_{n=1}^{m}\left\{1-\frac{1}{\left[1+\left(B_n\alpha\right)^{C_n}\right]^{\left[1-\left(1/C_n\right)\right]}}\right\}A_n \qquad [17]$$

where m is the number of regions where connecting pore throats are draining or filling, and A, B, C are constants. Figure 7–9 shows model (Eq. [17]) fits to gas permeability data for three Japanese soils from Poulsen and Moldrup (2006) using $n = 2$.

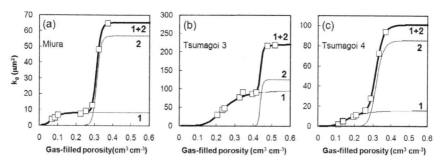

Fig. 7–9. Measured and predicted (Eq. [17]) ε–k_g relationships for three Japanese soils using $n = 2$. Modified from Poulsen and Moldrup (2006).

All three soils in Fig. 7–9 have systems of larger pores connected via smaller pore throats belonging to relatively narrow ranges of pore sizes. Values of the constants A, B, and C for each of the soils are shown in Table 7–2. Figure 7–9 also shows that the ε region where the drainage or filling of additional pores (the horizontal plateau in k_g) is widest for Miura (Fig. 7–9a), narrower for Tsumagoi 3 (Fig. 7–9b), and almost nonexistent for Tsumagoi 4 (Fig. 7–9c).

The value of the constant A is equal to the contribution to k_g by the corresponding term in Eq. [17]. In other words A_1 is equal to k_g at the first horizontal plateau in the ε–k_g curve, A_2 is equal to k_g at the second plateau minus k_g at the first plateau and so on. The constant B_1 is proportional to the value of ε at which the first term in Eq. [17] reaches 50% of its maximum value. The remaining constants have not yet been linked directly to soil physical properties.

The pore connectivity or organization index, PO, as defined by Eq. [5] for the three soils in Fig. 7–9 is shown in Fig. 7–10. When k_g increases rapidly with ε, PO also increases while in the regions where k_g is constant or only increases slowly with ε, PO actually decreases. Thus in addition to being an absolute measure of how well a pore system is connected PO can if plotted against ε also show where in the ε regime the pore system is getting connected. Thus, a plot of PO vs. ε could be used to more accurately identify the sizes of the pores that are critical in pore system connectivity.

Alternatively the slope of the ε–k_g relationship can as discussed earlier be used to represent the dynamics of pore connectivity as a function of ε. This is shown in Fig. 7–11 for the three Japanese soils from Fig. 7–9. In regions of ε where the critical connecting pore throats are being drained or filled the values of the slope $dk_g/d\varepsilon$ are high while in regions where non connecting pores are being drained $dk_g/d\varepsilon$ values are close to 0. For the three soils, the ε–$dk_g/d\varepsilon$ relationship exhibit distinct peaks in the regions of ε where connecting pore throats are being drained or filled. All three soils exhibit two peaks, a wider peak associated with the drainage or filling of the directly connected larger pores (at lower values of ε) and a more distinct, narrow peak associated with the drainage or filling of the

Table 7–2. Macro-pore and connecting pore diameters and values of the constants A, B, and C for the three soils in Fig. 9 ($n = 2$). Values for the constants were fitted using Eq. [17] by Poulsen and Moldrup (2006).

Soil	Macro-pores	Connecting pores	A_1	B_1	C_1	A_2	B_2	C_2
	——————— μm ———————		μm²			μm²		
Miura	>47	<9	8.3	16	3.8	57	3.2	29
Tsumagoi 3	>30	1–3	97	4.3	5.1	124	2.2	91
Tsumagoi 4	>9	<9	16	6.4	4.3	85	3.2	17

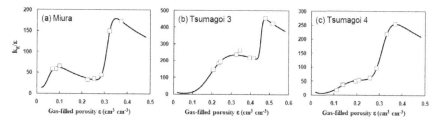

Fig. 7–10. Measured and predicted values of pore organization index (as defined by (Eq. [5]) for the three soils from Fig. 7–9. Predictions of PO were calculated using Eq. [17]. Data from Poulsen and Moldrup (2006).

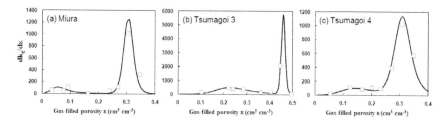

Fig. 7–11. Measured and predicted values of $dk_g/d\varepsilon$ for the three soils from Fig. 7–9. Predictions were calculated using Eq. [17]. Data from Poulsen and Moldrup (2006).

critical, pore throats that connects the remaining larger pores. The data in Fig. 7–11 were used together with measurements of soil water retention for the three soils to assess the sizes of both the directly connected larger pores and the critical connecting pore throats. These are shown in Table 7–2.

Gas Permeability in Soils with an Unstable Pore System
Variable Pore Connectivity and Pore Size Distribution
Variable pore size distribution and pore system connectivity are usually associated with soils that shrink or swell in response to variations in soil water content. This is especially the case for clayey soils that often show extensive crack formation under dry conditions. Variable pore size distribution and pore system connectivity is also observed in soils that are unstable due to recent disturbance, for instance via recent tilling (Slowinska-Jurkiewicz, 1994). In such soils wetting–drying or freezing–thawing cycles will cause the soil to move toward a more stable pore system geometry over time. In soils with variable pore system geometry, k_g can vary over several orders of magnitude as a function of time in response to changes in soil pore system geometry (Wells et al., 2007). Seguel and Horn (2006) investigated soil aggregation in volcanic soils with 13–28% clay size fraction in response to wetting and drying and found that aggregate strength increased with increasing pF meaning that dry aggregates are stron-

ger than wet ones. These authors also observed that aggregates increased their bulk density and decreased both porosity and pore size with increasing number of wetting-drying cycles. Also for some of the soils large changes in gas permeability was observed as a function of changes in water or air content. In general it has been extensively observed that for soils that have once been disturbed, wetting and drying cycles tend to increase aggregate strength over time and thereby increase the stability of the pore system (Seguel and Horn, 2006). This means that many soils with an unstable pore system created by a disturbance of the soil for instance by tilling, if left alone, often will progress toward a more stable pore system over time in response to climatic variations. An exception is clayey soils that also swell and shrink in response to variations in water content, however, these soils will very often not progress to a stable pore system but continue to have an unstable pore system. Poulsen et al. (2008) investigated gas permeability in soils with variable structure as a function of water content and found that in such soils gas permeability is not necessarily a monotonic increasing function of gas-filled porosity but may in certain cases decrease with increasing gas-filled porosity (drying) due to destruction of aggregates. This was especially observed in more coarse textured (silty and sandy soils).

Gas Permeability–Gas-Filled Porosity Relations in Soils with an Unstable Pore System

Poulsen et al. (2008) investigated gas permeability for a range of soils ranging from sand to silt including also composts and peat sols with clay contents below 5%. The materials were wetted to selected water content and then heavily disturbed (mixed) to produce structure. It turned out that for all materials k_g increased strongly with ε at low ε but reached a maximum at a certain ε value after which k_g decreased with increasing ε at intermediate ε which is in contradiction to most observations of the k_g–ε relationship in unsaturated porous media. At high ε (dry soils) materials again showed increasing k_g for increasing ε. Figure 7–12 shows examples of k_g–ε relationships for three soils from Poulsen et al. (2008).

In all three cases it is seen that $\log(k_g)$ varies with $\log(\varepsilon)$ in a strongly nonlinear manner. The physical reason for this is that at low values of ε, the soils are non-aggregated and $\log(k_g)$ increases with $\log(\varepsilon)$ both because the pore system is draining and because the soils become progressively more aggregated as the water content is reduced. This process corresponds to the behavior of many soils that shrink and swell in response to water content changes (such as clayey soils). As ε increases, k_g for the soils in Fig. 7–12 reaches a local maximum value after which it begins to decrease with increasing ε. The reason here is that the aggregates formed for the soils in Fig. 7–12 are rather weak and therefore will tend to disintegrate when the soil becomes

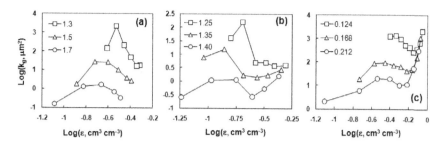

Fig. 7–12. Measured log(k_g)–log(ε) relationships for three soils with unstable pore systems, (a) Hjørring silty sand, (b) Foulum sandy loam, (c) peat. Numbers on figures refer to soil dry bulk density. Modified from Poulsen et al. (2008).

drier (as these soils are sandy). This causes a decrease in k_g with increasing ε as the inter-aggregate porosity (consisting of larger pores) decreases with the destruction of the aggregates. This effect does not take place in clayey soils where the aggregates formed are stronger and therefore of a more permanent character, but in more sandy soils that are being disturbed for instance due to tilling this effect is likely to take place. For high ε, k_g again increases with ε as the soil dries out and the gas-filled pore system becomes better connected once again. The conceptual relationship between log(k_g) and log(ε) is illustrated in Fig. 7–13.

The log(k_g)–log(ε) relationship can be divided into three regions (Poulsen et al., 2008) with respect to the behavior of log (k_g) in response to changes in log(ε). In region I, the soil aggregation and formation of large inter-aggregate pores (cracks) increase with increasing ε for instance due to shrinking (in clayey soils) or due to formation of weak aggregate structures (in silty soils). Clayey soils will in general only exhibit log(k_g)–log(ε) relationships in this region as the aggregates, and inter-aggregate pores (cracks) formed in these soils are usually stable also at high values of ε. In region II, weak aggregates or weak structure tends to be destroyed

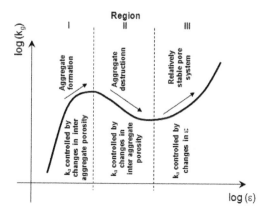

Fig. 7–13. Conceptual relationship between log(k_g)–log(ε) in soils with unstable pore systems and variable structure forming potentials in response to changes in water content (Poulsen et al. 2008).

with increasing ε and the inter-aggregate pore network gradually disappear in favor of smaller pores. This mechanism results in decreasing k_g with increasing ε. At present only very few measurements have been made under conditions where this mechanism take place in soils and knowledge about it is therefore limited, however, present data indicate that it may take place in coarser soils (such as silty and loamy soils) which are being disturbed due to tilling, freeze–thaw cycles or other types of processes that results in a disturbance of the soil. In region III, the soil pore system is becoming stable as weak aggregates and pore structures have been destroyed and k_g once again increases with increasing ε just as observed for soils with stable pore systems. It could be argued that soils with a well connected stable pore system will exhibit ε–k_g relationships that belong to region III. In general it is so that not all soils with unstable pore systems exhibit ε–k_g relationships that cover all three regions. For the majority of soils the ε–k_g relationship will only cover part of the three regions such as is illustrated in Fig. 7–14 where the Hjørring soil (Fig. 7–12a) exhibits ε–k_g relationships belonging to region I and II, while both the Foulum and the peat soils (Fig. 7–12b and 7–12c) exhibit ε–k_g relationships covering all three regions (for some bulk densities).

Modeling Gas Permeability–Gas-Filled Porosity Relations in Soils with an Unstable Pore System

Poulsen and Blendstrup (2008) found for the same six materials investigated by Poulsen et al. (2008) that even though the relationship between $\log(k_g)$ and $\log(ε)$ for fixed values of dry bulk density, ρ_b (such as those shown in Fig. 7–12), is strongly nonlinear, the relationship between $\log(k_g)$ and $\log(ε)$ for constant values of gravimetric water content, ω, (g/g soil) but varying values of ρ_b, is generally linear. Figure 7–14 shows examples of $\log(k_g)$–$\log(ε)$ relationships for constant values of ω for the three soils in Fig. 7–12. In all cases the $\log(k_g)$–$\log(ε)$ relationship is with good approximation linear. Figure 7–14 also show that there is a tendency

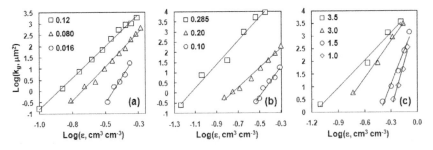

Fig. 7–14. Measured relationships between $\log(k_g)$–$\log(ε)$ for constant values of gravimetric water content ω (g H_2O g soil^{-1}) for the Hjørring silty sand, Foulum loamy sand and Peat soils in Fig. 7–12. Lines are best fit linear relationships to the measured data. Modified from Poulsen and Blendstrup (2008).

for the slope of the $\log(k_g)$–$\log(\varepsilon)$ relationship to be higher at lower values of ω. This was also observed for the other materials investigated by Poulsen and Blendstrup (2008). The linear relationship between $\log(k_g)$ and $\log(\varepsilon)$ was described by Poulsen and Blendstrup (2008) as

$$\log(k_g) = A\,[\log(\varepsilon) - \log(\varepsilon_{th})] \tag{18}$$

where A is the slope of the $\log(k_g)$–$\log(\varepsilon)$ relationship and ε_{th} is the value of ε at which $\log(k_g)$ is zero, that is, $k_g = 1\ \mu m^2$. The increasing values of A with ω suggests that A and ε_{th} are related. Poulsen and Blendstrup (2008) discovered a linear relationship between $\log(A)$ and ε_{th}. This relationship which is illustrated in Fig. 7–15 for the soils from Fig. 7–12 is described as

$$\log(A) = \alpha\,\varepsilon_{th} + \beta \tag{19}$$

where α is the slope of the ε_{th}–$\log(A)$ relationship and β is the intercept with the $\log(A)$ axis. Figure 7–15 shows that the ε_{th}–$\log(A)$ relationships for the three soils in general are linear with some minor variations. It was observed (Poulsen and Blendstrup, 2008) that there was no direct relationship between soil texture and the value of α. Combining Eq. [18] and [19] yields:

$$\log\!\left(k_g\right) = 10^{\left(\alpha\varepsilon_{th} + \beta\right)}\log\!\left(\varepsilon/\varepsilon_{th}\right) \tag{20}$$

Because A varies with ω (Fig. 7–14) and also with ε_{th} (Fig. 7–15) it means that ω and ε_{th} are also related. Poulsen and Blendstrup (2008) found that ε_{th} and ω were linearly related. This is shown in Fig. 7–16 for the three soils from Fig. 7–12. In all three cases the ε_{th}–ω relationship is linear with good approximation. At high ω (near saturation) the linear relationship fails to describe the ε_{th}–ω relationship, likely because k_g is very uncertain and difficult to determine accurately near

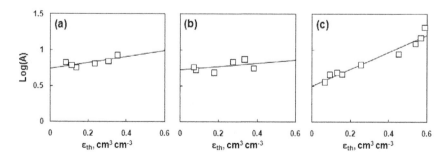

Fig. 7–15. Measured relationships between $\log(A)$ and ε_{th} for (a) Hjørring silty sand, (b) Foulum loamy sand, and (c) peat soils in Fig. 7–12. Lines are best fit linear relationships to the measured data. Modified from Poulsen and Blendstrup (2008).

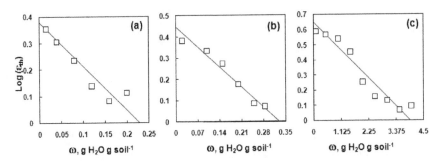

Fig. 7–16. Measured relationships between ω and ε_{th} for (a) Hjørring silty sand, (b) Foulum loamy sand, and (c) peat soils in Fig. 7–12. Lines are best fit linear relationships to the measured data. Modified from Poulsen and Blendstrup (2008).

saturation. This is seen in Fig. 7–16a and 7–16c. The linear relationship between ε_{th} and ω can be described as:

$$\varepsilon_{th} = \gamma \omega + \delta \tag{21}$$

where γ is the slope of the $\omega-\varepsilon_{th}$ relationship and d is the intercept, that is, at $\omega = 0$. Combining Eq. [20] and [21] yields

$$\log\left(k_g\right) = 10^{\left(\alpha\varepsilon_{th}+\beta\right)}\log\left[\varepsilon/\left(\gamma\omega+\delta\right)\right] \tag{22}$$

This equation may alternatively be written in terms of ω and ρ_b instead of ω and ε as

$$\log\left(k_g\right) = 10^{\left(\alpha\varepsilon_{th}+\beta\right)}\log\left[\frac{1-\rho_b\omega-\left(\rho_b/\rho_s\right)}{\gamma\omega+\delta}\right] \tag{23}$$

Poulsen and Blendstrup (2008) proposed that it would be possible to predict k_g as a function of ε and ω (or ρ_b and ω) for a given soil based on a minimum of four measurements of k_g at known values of ε and ω (or ρ_b and ω). The approach is illustrated in Fig. 7–17 and is as follows. Initially two values of ω, ω_1, and ω_2, representing the ω range for the soil across all ε (or ρ_b) values are selected. It is important that the ω values are selected such that they can be achieved for a relatively wide range of ε (or ρ_b) to get the best estimates of the slope A. It is also important not to select points that are too close to saturated conditions to avoid uncertainties in the k_g determination. Two soil samples with $\omega = \omega_1$ and two samples with $\omega = \omega_2$ are prepared such that they cover the range of possible ε at each ω. Again it is important to remember that the larger the $k_g-\varepsilon$ area spanned by the four (ω, ε) points represented by the samples, the more accurate the estimates of k_g. For each of the four samples k_g is measured and the slope, A, and intercept ε_{th}

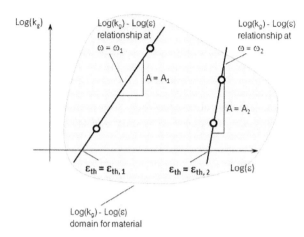

Fig. 7–17. Approach for predicting k_g as a function of ε and ω based on four measurements of k_g at selected ε, ω values. Modified from Poulsen and Blendstrup (2008).

of the $\log(k_g)$–$\log(\varepsilon)$ relationship are determined for ω_1 and ω_2 [two points for each $\log(k_g)$–$\log(\varepsilon)$ relationship] using Eq. [17] yielding values A_1 and A_2. Estimates of parameters α and β are then determined from the two A values using Eq. [19] and parameters γ and δ are determined using the two values of ε_{th} together with Eq. [21]. The values of α, β, γ and δ can now be inserted in Eq. [22] or [23] to predict k_g as a function of ε in the entire k_g–ε domain. It is important to bear in mind that although Eq. [22] and [23] can be evaluated for ε, ω, and ρ_b values outside the range possible for the soil, the k_g estimates produced by such calculations will not have any physical meaning. It is therefore important to have some knowledge about the k_g–ε range for the soil when interpreting k_g predictions.

The concept illustrated in Fig. 7–17 was tested against measurements for different materials (Poulsen and Blendstrup, 2008) and found robust enough to yield acceptable predictions of k_g as a function of ε for variable ω. In general predictions of k_g were within one-half order of magnitude from the measured values. Figure 7–18 shows measured and predicted values of k_g as a function of ε for the three soils in Fig. 7–12.

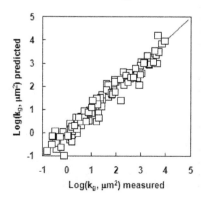

Fig. 7–18. Measured and predicted values of $\log(k_g)$ using the prediction approach in Fig. 7–17 for the three soils from Fig. 7–12. Modified from Poulsen and Blendstrup (2008).

It is seen that the approach is able to predict $\log(k_g)$ with good approximation. This was also observed for the other materials tested by Poulsen and Blendstrup (2008). The most important issue when using the approach is to properly select the four combinations of ω and ε (Fig. 7–17) so they cover the $k_g-\varepsilon$ domain for the soil in question without moving into the area close to saturated conditions. This can especially for intact soil samples collected in the field present somewhat of a challenge as the possible ranges of ω and ε are usually not known before collecting the samples.

Summary

Existing knowledge about characterization of soil pore system characteristics using the relationship between gas permeability (k_g) and gas-filled porosity (ε) was discussed based on existing literature in the field. Soils could generally be divided into three types with respect to pore system characteristics. (i) Soils with a stable and well connected pore system where (almost) all pores participate in conducting fluid; (ii) soils with a stable but poorly connected pore system where only a limited fraction of the pores participate in conducting fluid; and (iii) soils with an unstable pore system where variable fractions of the pores conduct fluid depending on soil structure.

Soils with a stable and well-connected pore system often have a uni-modal pore size distribution. In such soils the $k_g-\varepsilon$ relationship is usually a monotonous increasing (power) function and very often it is linear in a $\log(k_g)-\log(\varepsilon)$ system. Pore organization indices for such soils represented by k_g/ε and $dk_g/d\varepsilon$ as a function of ε are also both monotonous increasing (power) functions. Observations indicate that at least for coarse textured media (sand and compost) the $\log(k_g)-\log(\varepsilon)$ relationships for materials of similar texture are not only linear but they intercept each other in the same $\log(k_g)-\log(\varepsilon)$ point allowing for estimation of the $k_g-\varepsilon$ relationship based on a single $k_g-\varepsilon$ measurement at some distance from the common interception point.

Soils with stable but poorly connected pore systems can be divided into two sub-categories as (a) soils with a limited but well connected pore system surrounded by small or poorly connected pores, (b) soils with a system of larger pores connected via small pore throats surrounded by small or poorly connected pores. If the surrounding pores are of the same sizes as the connected ones the soil will likely have a uni-modal pore-size distribution while if the surrounding pores are smaller the soil likely has a bimodal or multi-modal pore-size distribution. In these soils the $k_g-\varepsilon$ relationship consists of sections where the relationship is monotonously increasing (here connecting pores are filling or draining changing the connectivity of the pore system) with intermittent plateaus (here mainly un-connecting pores are being drained or filled). The k_g/ε–ε relationship will consist of sections with increasing values (here the pore system is getting increasingly connected) and decreasing values (here the pore system connectivity is constant as no new pore connections are being made). The $dk_g/d\varepsilon$–ε relationship will exhibit peaks in the regions where the pore system is getting better connected (increasing connectivity) and values near zero when no changes in pore system connectivity take place. In such soils an empirical concept for predicting the k_g–ε relationship is available but the empirical constants needed to evaluate the relationship are not yet known and no existing procedure is available for obtaining them other than measuring the entire k_g–ε relationship for the soil in question which will of course render the prediction unnecessary.

Soils with an unstable pore system are often characterized by changes in aggregate formation and destruction in response to changes in water content. In these soils the fractions of inter and intra-aggregate porosity will change strongly with changes in water content. The k_g–ε relationship for these soils is strongly nonlinear with sections of increasing and decreasing k_g values. Observations indicate that in case of constant dry bulk density (ρ_b) and total porosity (ϕ), the ε-domain can be divided into three regions with respect to pore system behavior. At high water contents (region I) soil aggregation increases in response to increasing ε (decreasing water content), causing k_g to increase rapidly. At intermediate water contents (region II) aggregate stability decreases and the aggregates are increasingly destroyed as ε increases. This causes k_g to decrease with increasing ε in this region. At low water contents (region III) unstable aggregates have been destroyed and the soil now behaves as a soil with a stable pore and often well connected pore system where k_g increases monotonically with increasing ε. In these soils k_g/ε and $dk_g/d\varepsilon$ are also strongly nonlinear. Observations, however indicated that for fixed gravimetric water content (variable ρ_b and ϕ) the $\log(k_g)$–$\log(\varepsilon)$ relationship is linear. Based on this, an approach for predicting k_g as a function of ε in the entire k_g–ε domain for a given soil or other porous material based on four k_g–ε measurements has been developed.

References

Ahuja, L.R., J.W. Naney, R.E. Green, and D.R. Nielsen. 1984. Macroporosity to characterize spatial variability of hydraulic conductivity and effects of land management. Soil Sci. Soc. Am. J. 48:699–702. doi:10.2136/sssaj1984.03615995004800040001x.

Alexander, L., and R.W. Skaggs. 1986. Predicting unsaturated hydraulic conductivity from the soil water characteristic. Trans. ASAE 29:176–184.

Ball, B.C. 1981. Modelling of soil pores as tubes using gas permeabilities, gas diffusivities and water release. J. Soil Sci. 32:465–481. doi:10.1111/j.1365-2389.1981.tb01723.x.

Ball, B.C., M.F. Sullivan, and R. Hunter. 1988. Gas diffusion, fluid flow and derived pore continuity indices in relation to vehicle traffic and tillage. J. Soil Sci. 39:327–339. doi:10.1111/j.1365-2389.1988.tb01219.x.

Blackwell, P.S., A.J. Ringrose-Voase, N.S. Jayawardane, K.A. Olsson, and D.C. Mason. 1990. The use of air-filled porosity and intrinsic permeability to characterize structure of macropore space and saturated hydraulic conductivity of clay soils. J. Soil Sci. 41:215–228. doi:10.1111/j.1365-2389.1990.tb00058.x.

Brutsaert, W. 1966. Probability laws for pore size distributions. Soil Sci. 101:85–92. doi:10.1097/00010694-196602000-00002.

Buehrer, T.F. 1932. The movement of gases through the soil as a criterion of soil structure. Tech. Bull. 39, Ariz. Agric. Exp. Station, Univ. Ariz., Tucson, AZ.

Campbell, G.S. 1974. A simple method for determining unsaturated conductivity from moisture retention data. Soil Sci. 117:311–314. doi:10.1097/00010694-197406000-00001.

Clapp, R.B., and G.M. Hornberger. 1978. Empirical equations for some soil hydraulic properties. Water Resour. Res. 14:601–604. doi:10.1029/WR014i004p00601.

Corder, G.W., and D.I. Forman. 2009. Non-parametric statistics for non-statisticians: A step by step approach. John Wiley and Sons, Hoboken, NJ.

Fish, A.N., and A.J. Koppi. 1993. The use of a simple field air permeameter as a rapid indicator of functional soil pore space. Geoderma 63:255–264.

Freeze, R.A. 1994. Henry Darcy and the fountains of Dijon. Ground Water 32:23–30. doi:10.1111/j.1745-6584.1994.tb00606.x.

Gostomski, P., and L. Liaw. 2001. Air permeability of biofilter media. Session AE-2a. Proceedings of Air and Waste Management Assoc. 94th Annual Meeting and Exhibition, Orlando, FL, June 18–24.

Green, R.D., and S.J. Fordham. 1975. A field method for determining air permeability in soil. Ministry of Agriculture, Fisheries and Food. Tech. Bull. 29, Soil physical conditions and crop production. Her Majesty's Stationary Off., London.

Gronenvelt, P.H., B.D. Kay, and C.D. Grant. 1984. Physical assessment of soil with respect to rooting potential. Geoderma 34:101–114. doi:10.1016/0016-7061(84)90016-8.

Grover, B.L. 1955. Simplified air permeameters for soil in place. Soil Sci. Soc. Am. Proc. 19:414–418. doi:10.2136/sssaj1955.03615995001900040006x.

Iversen, B.V., P. Schjønning, T.G. Poulsen, and P. Moldrup. 2001. In situ, on site and laboratory measurements of soil air permeability: Boundary conditions and measurement scale. Soil Sci. 166:97–106. doi:10.1097/00010694-200102000-00003.

Kawamoto, K., P. Moldrup, P. Schjonning, B.V. Iversen, T. Komatsu, and D.E. Rolston. 2006. Gas transport parameters in the vadose zone: Development and tests of power-law models for air permeability. Vadose Zone J. 5:1205–1215. doi:10.2136/vzj2006.0030.

Kirkham, D., M. De Boodt, and L. De Leenheer. 1958. Air permeability at the field capacity as related to soil structure and yields. In: Proceedings of the international symposium on soil structure. Rijkslandbouwhogeschool, Ghent, Belgium. p. 377–391.

Loll, P., P. Moldrup, P. Schjønning, and H. Riley. 1999. Preducting saturated hydraulic conductivity from air permeability: Application in stochastic water infiltration modelling. Water Resour. Res. 35:2387–2400. doi:10.1029/1999WR900137.

McKenzie, B.M., and A.R. Dexter. 1996. Methods for studying the permeability of individual soil aggregates. J. Agric. Eng. Res. 65:23–28. doi:10.1006/jaer.1996.0076.

Moldrup, P., T. Olesen, T. Komatsu, P. Schjønning, and D.E. Rolston. 2001. Tortuosity, diffusivity and permeability in the soil liquid and gaseous phases. Soil Sci. Soc. Am. J. 65:613–623. doi:10.2136/sssaj2001.653613x.

Moldrup, P., T.G. Poulsen, P. Schjønning, T. Olesen, and T. Yamaguchi. 1998. Gas permeability in undisturbed soils: Measurements and predictive models. Soil Sci. 163:180–189. doi:10.1097/00010694-199803000-00002.

Moldrup, P., S. Yoshikawa, T. Olesen, T. Yamaguchi, and D.E. Rolston. 2003. Air permeability in undisturbed volcanic ash soils: Predictive model test and soil structure fingerprint. Soil Sci. Soc. Am. J. 67:32–40. doi:10.2136/sssaj2003.0032.

Poulsen, T.G., and H. Blendstrup. 2008. Predicting air permeability in porous media with variable structure, bulk density and water content. Vadose Zone J. 7:1223–1229.

Poulsen, T.G., H. Blendstrup, and P. Schjønning. 2008. Air permeability in porous media with variable structure forming potential. Vadose Zone J. 7:1139–1143. doi:10.2136/vzj2007.0147.

Poulsen, T.G., B.V. Iversen, T. Yamaguchi, P. Moldrup, and P. Schjønning. 2001. Spatial and temporal dynamics of air permeability in a constructed, agricultural field. Soil Sci. 166:153–162. doi:10.1097/00010694-200103000-00001.

Poulsen, T.G., P. Moldrup, P. Schjonning, and J. Aa. Hansen. 1998. Gas permeability and diffusivity in undisturbed soils: SVE implications. J. Environ. Eng. 124:979–986. doi:10.1061/(ASCE)0733-9372(1998)124:10(979).

Poulsen, T.G., and P. Moldrup. 2006. Bi-modal probability law model for unified description of water retention, air and water permeability and gas diffusivity in variably saturated soil. Vadose Zone J. 5:1119–1128. doi:10.2136/vzj2005.0146.

Poulsen, T.G., and P. Moldrup. 2007. Air permeability of compost as related to bulk density and volumetric air content. Waste Manage. Res. 25:343–351. doi:10.1177/0734242X07077819.

Poulsen, T.G., P. Moldrup, and P. Schjønning. 2007. Predicting air permeability in sandy soils as a function of volumetric air content using one reference-point value of air permeability. J. Environ. Eng. 133:995–1001. doi:10.1061/(ASCE)0733-9372(2007)133:10(995).

Resurreccion, A.C., K. Kawamoto, T. Komatsu, P. Moldrup, K. Sato, and D.E. Rolston. 2007. Gas diffusivity and air permeability in a volcanic ash soil profile: Effects of organic matter and water retention. Soil Sci. 172:432–443. doi:10.1097/SS.0b013e3180471c94.

Saxton, K.E., J.F. Kenny, and D.K. McCool. 1993. Air permeability to define frozen soil infiltration with variable tillage and residue. Trans. ASAE 36:1369–1375.

Schjønning, P. 1986. Soil permeability by air as influenced by soil type and incorporation of straw (in Danish). Tidsskr. Planteavl 90:227–240.

Schjønning, P., B.V. Iversen, L.J. Munkholm, R. Labouriau, and O.H. Jacobsen. 2005. Pore characteristics and hydraulic properties of a sandy loam supplied for a century with either animal manure or mineral fertilizers. Soil Use Manage. 21:265–275.

Schjønning, P., L.J. Munkholm, P. Moldrup, and O.H. Jacobsen. 2002. Modelling soil pore characteristics from measurements of air exchange: The long-term effects of fertilization and crop rotation. Eur. J. Soil Sci. 53:331–339. doi:10.1046/j.1365-2389.2002.00438.x.

Seguel, O., and R. Horn. 2006. Structure properties and pore dynamics in aggregate beds due to wetting-drying cycles. J. Plant Nutrition Soil Sci. 169:221–232. doi:10.1002/jpln.200521854.

Slowinska-Jurkiewicz, A. 1994. Changes in the structure and physical properties of soil during spring tillage operations. Soil Tillage Res. 29:397–407. doi:10.1016/0167-1987(94)90111-2.

Tuli, A., J.W. Hopmans, D.E. Rolston, and P. Moldrup. 2005. Comparison of air and water permeability between disturbed and undisturbed soils. Soil Sci. Soc. Am. J. 69:1361–1371. doi:10.2136/sssaj2004.0332.

van Genuchten, M.Th. 1980. A closed-form equation for predicting the hydraulic conductivity of unsaturated soils. Soil Sci. Soc. Am. J. 44:892–898. doi:10.2136/sssaj1980.03615995004400050002x.

Wells, T., S. Fityus, and D.W. Smith. 2007. Use of in situ air flow measurements to study permeability in cracked clay soils. J. Geotech. Geoenviron. Eng. 133:1577–1586. doi:10.1061/(ASCE)1090-0241(2007)133:12(1577).

Page numbers followed by f and t indicate figures and tables.

Advances in Agricultural Systems Modeling
Transdisciplinary Research, Synthesis, and Applications

Agriculture today is complicated by growing environmental concerns, increasingly limited water available for agriculture, market-based global competition that challenges traditional production systems, more frequent droughts and climate change, and the growing production of bioenergy crops. Whole system–based quantitative planning and decision tools are needed to guide the optimum management of resources while addressing these problems. Process-level models of agricultural systems, integrated with focused field research, are required to develop these tools and extend their application to different soils, climates, and situations.

Agricultural system modeling has made substantial progress, but there are still many critical gaps in our knowledge of various processes, and especially their interactions. Most of these gaps occur at the boundaries of disciplines, and further transdisciplinary research is needed. There is also the need for better synthesis and quantification of knowledge at the whole system level, both to improve system models and realize a more systematic and collaborative approach to building models of the future. At the same time, we need to facilitate application of these models to address and solve current real-world problems. For this purpose, we need to understand the different scales of application and the scale-dependence of parameters.

The American Society of Agronomy, Crop Science Society of America, and Soil Science Society of America are taking a leadership role in encouraging transdisciplinary and interdisciplinary research and its synthesis to solve practical problems. We believe that the future breakthroughs in science and technology lie, indeed, at the boundaries of disciplines. Therefore, the Societies have initiated the new series Advances in Agricultural Systems Modeling, A Series of Transdisciplinary Research, Synthesis, and Applications.

The purpose of the series is to:

- Encourage and advance critical transdisciplinary research, and its synthesis and quantification, through publication of anonymously peer-reviewed papers by top researchers worldwide in a given knowledge gap area. The writing of papers and publication will generally be preceded by a workshop of these researchers, where the papers will be presented and discussed. The papers will contain both new research and new or improved concepts for synthesis.

- Encourage collaboration among the top researchers in new research, synthesis, and building and improvement of model components. As often as possible and where appropriate, the authors will be asked to provide computer code of the components to be shared by researchers and modelers.

- Encourage and advance the application of system models to solve practical problems through publishing case studies of such applications, along with illustrated instructions on how the models are used. For example, an application may be the use of particular models to optimize the water and nutrients under limited water conditions and to evaluate the effects of climate change on agriculture.

- Encourage better instruction in these models and their application. The state-of-the-science syntheses given by each author will be highly useful for both undergraduate and graduate teaching. The new research in the papers and model component codes will be valuable for graduate level teaching, research, and training of student in the use of models.

Laj R. Ahuja, Series Editor

Printed and bound by CPI Group (UK) Ltd, Croydon, CR0 4YY

27/10/2024

14580266-0001